INTERNATIONAL
WILDLIFE
ENCYCLOPEDIA

THIRD EDITION

Volume 11
LEO–MAR

Marshall Cavendish Corporation
99 White Plains Road
Tarrytown, New York 10591–9001

Website: www.marshallcavendish.com

© 2002 Marshall Cavendish Corporation

Library of Congress Cataloging-in-Publication Data

Burton, Maurice, 1898-
    International wildlife encyclopedia / [Maurice Burton, Robert Burton] .-- 3rd ed.
        p. cm.
    Includes bibliographical references (p.).
    Contents: v. 1. Aardvark - barnacle goose -- v. 2. Barn owl - brow-antlered deer -- v. 3. Brown bear - cheetah -- v. 4. Chickaree - crabs -- v. 5. Crab spider - ducks and geese -- v. 6. Dugong - flounder -- v. 7. Flowerpecker - golden mole -- v. 8. Golden oriole - hartebeest -- v. 9. Harvesting ant - jackal -- v. 10. Jackdaw - lemur -- v. 11. Leopard - marten -- v. 12. Martial eagle - needlefish -- v. 13. Newt - paradise fish -- v. 14. Paradoxical frog - poorwill -- v. 15. Porbeagle - rice rat -- v. 16. Rifleman - sea slug -- v. 17. Sea snake - sole -- v. 18. Solenodon - swan -- v. 19. Sweetfish - tree snake -- v. 20. Tree squirrel - water spider -- v. 21. Water vole - zorille -- v. 22. Index volume.
    ISBN 0-7614-7266-5 (set) -- ISBN 0-7614-7267-3 (v. 1) -- ISBN 0-7614-7268-1 (v. 2) -- ISBN 0-7614-7269-X (v. 3) -- ISBN 0-7614-7270-3 (v. 4) -- ISBN 0-7614-7271-1 (v. 5) -- ISBN 0-7614-7272-X (v. 6) -- ISBN 0-7614-7273-8 (v. 7) -- ISBN 0-7614-7274-6 (v. 8) -- ISBN 0-7614-7275-4 (v. 9) -- ISBN 0-7614-7276-2 (v. 10) -- ISBN 0-7614-7277-0 (v. 11) -- ISBN 0-7614-7278-9 (v. 12) -- ISBN 0-7614-7279-7 (v. 13) -- ISBN 0-7614-7280-0 (v. 14) -- ISBN 0-7614-7281-9 (v. 15) -- ISBN 0-7614-7282-7 (v. 16) -- ISBN 0-7614-7283-5 (v. 17) -- ISBN 0-7614-7284-3 (v. 18) -- ISBN 0-7614-7285-1 (v. 19) -- ISBN 0-7614-7286-X (v. 20) -- ISBN 0-7614-7287-8 (v. 21) -- ISBN 0-7614-7288-6 (v. 22)
    1. Zoology -- Dictionaries.   I. Burton, Robert, 1941-   . II. Title.

    QL9 .B796 2002
    590'.3--dc21

                                                        2001017458

Printed in Malaysia
Bound in the United States of America

07 06 05 04 03 02 01   8 7 6 5 4 3 2 1

Brown Partworks
Project editor: Ben Hoare
Associate editors: Lesley Campbell-Wright, Rob Dimery, Robert Houston, Jane Lanigan, Sally McFall, Chris Marshall, Paul Thompson, Matthew D. S. Turner
Managing editor: Tim Cooke
Designer: Paul Griffin
Picture researchers: Brenda Clynch, Becky Cox
Illustrators: Ian Lycett, Catherine Ward
Indexer: Kay Ollerenshaw

Marshall Cavendish Corporation
Editorial director: Paul Bernabeo

Authors and Consultants

Dr. Roger Avery, BSc, PhD (University of Bristol)

Rob Cave, BA (University of Plymouth)

Fergus Collins, BA (University of Liverpool)

Dr. Julia J. Day, BSc (University of Bristol), PhD (University of London)

Tom Day, BA, MA (University of Cambridge), MSc (University of Southampton)

Bridget Giles, BA (University of London)

Leon Gray, BSc (University of London)

Tim Harris, BSc (University of Reading)

Richard Hoey, BSc, MPhil (University of Manchester), MSc (University of London)

Dr. Terry J. Holt, BSc, PhD (University of Liverpool)

Dr. Robert D. Houston, BA, MA (University of Oxford), PhD (University of Bristol)

Steve Hurley, BSc (University of London), MRes (University of York)

Tom Jackson, BSc (University of Bristol)

E. Vicky Jenkins, BSc (University of Edinburgh), MSc (University of Aberdeen)

Dr. Jamie McDonald, BSc (University of York), PhD (University of Birmingham)

Dr. Robbie A. McDonald, BSc (University of St. Andrews), PhD (University of Bristol)

Dr. James W. R. Martin, BSc (University of Leeds), PhD (University of Bristol)

Dr. Tabetha Newman, BSc, PhD (University of Bristol)

Dr. J. Pimenta, BSc (University of London), PhD (University of Bristol)

Dr. Kieren Pitts, BSc, MSc (University of Exeter), PhD (University of Bristol)

Dr. Stephen J. Rossiter, BSc (University of Sussex), PhD (University of Bristol)

Dr. Sugoto Roy, PhD (University of Bristol)

Dr. Adrian Seymour, BSc, PhD (University of Bristol)

Dr. Salma H. A. Shalla, BSc, MSc, PhD (Suez Canal University, Egypt)

Dr. S. Stefanni, PhD (University of Bristol)

Steve Swaby, BA (University of Exeter)

Matthew D. S. Turner, BA (University of Loughborough), FZSL (Fellow of the Zoological Society of London)

Alastair Ward, BSc (University of Glasgow), MRes (University of York)

Dr. Michael J. Weedon, BSc, MSc, PhD (University of Bristol)

Alwyne Wheeler, former Head of the Fish Section, Natural History Museum, London

Picture Credits

Contents

LEOPARD

A Javan leopard, Indonesia. In Southeast Asia, the leopard's forest habitats have been destroyed on a massive scale by logging. Java has lost over 90 percent of its original forest so that now only a few hundred Javan leopards remain in the wild.

THE SIZE OF THE LEOPARD, a close relative of the lion and tiger, varies from one part of its range to another. It is found in southern Asia through Central Asia and India into China, in parts of Arabia and in sub-Saharan and northeastern Africa. The male may be more than 6 feet (1.8 m) long, with a 3-foot (90-cm) tail, and can weigh up to 200 pounds (90 kg). The female is smaller and rarely exceeds 130 pounds (60 kg).

The color and length of the leopard's fur also vary with geographical location and with climate. Its ground color is from pale yellow or buff to tawny yellow or deep chestnut, the underparts being paler or whitish. There are many small, black spots that are arranged in rosettes on most of the body. As a result of these differences in size and appearance, about 25 subspecies or races of leopard have been named. Leopards are also called panthers, with this name often used to distinguish the few individuals that are completely melanistic (totally black).

Powerful cat

Always shy and wary, with keen senses, the leopard's ability to hide makes it harder to track down than a lion or tiger. Leopards live wherever cover is available: in forest, bush and scrub or on rocky hillsides. They are mainly solitary except during the breeding season. Leopards tend to be nocturnal when living in areas where they are hunted or persecuted by humans. Elsewhere they are active in the early morning and again in the late afternoon, continuing into the night. During much of the day, leopards rest in "daybeds" in thick undergrowth.

Leopards are extremely powerful relative to their size and can easily drag a large kill up into a tree. They climb trees well and often take a kill up into a fork to cache (store) it out of reach from other predators, such as lions and hyenas. Leopards can also leap long distances, perhaps up to 20 feet (6 m). Their voice is a grunting or harsh coughing, or a sawing roar.

LEOPARD

CLASS	**Mammalia**
ORDER	**Carnivora**
FAMILY	**Felidae**
GENUS AND SPECIES	***Panthera pardus***

ALTERNATIVE NAME
Panther

WEIGHT
Male: 82–200 lb. (37–90 kg).
Female: 62–130 lb. (28–60 kg).

LENGTH
Head and body: 3–6¼ ft. (0.9–1.9 m);
shoulder height: 1½–2½ ft. (45–75 cm);
tail: 2–3½ ft. (0.6–1.1. m)

DISTINCTIVE FEATURES
Strong, muscular body; light yellow to buff
or deep chestnut coat with dark rosettes of
spots; paler underbelly; some individuals
completely melanistic (totally black)

DIET
Small to medium-sized grazing animals such
as wildebeest, antelopes, gazelles and deer;
also monkeys, warthogs, bushpigs, rodents,
hyraxes, porcupines, birds and invertebrates

BREEDING
Age at first breeding: 3 years; breeding
season: all year; number of young: usually
2 or 3; gestation period: 90–105 days;
breeding interval: 1–2 years

LIFE SPAN
Up to 23 years in captivity

HABITAT
Lowland forest, grassland, scrub, semidesert
and rocky hillsides

DISTRIBUTION
Much of Africa, through Middle East and
Central Asia to India, China and Indonesia

STATUS
Uncommon; several subspecies endangered

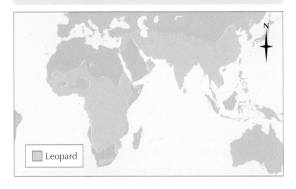

Leopard

Strong food preferences

The leopard feeds mainly on small to medium-sized grazing animals, such as wildebeest, deer and gazelles, but will eat almost anything that moves. Prey might range from dung beetles to antelopes much larger than itself. Commonly it preys upon impala, steenbok, reedbuck, zebra foals, wildebeest calves, warthogs and hyraxes. Smaller prey includes baboons and other monkeys, rodents such as rats and hares, porcupines, ground birds and arthropods. Small prey is taken more especially by young and old animals.

Several times it has been found that individual leopards develop a taste for one particular kind of prey. One leopard that was studied fed largely on impala, while another ate only bushpigs and would travel 2 miles (3 km) each night from its daybed to hunt these animals. It would seldom molest or take game closer to home.

Leopards use trees for cover and to store their kills away from other predators, for example lions.

Cubs are born blind

Breeding is likely to take place throughout the year, but with peaks of activity at different times depending on geographical location. Gestation is 90–105 days, and the cubs are born in a secluded cave, thicket or hollow tree. The number of cubs is usually two or three in a litter, although the range is between one and six. The cubs are born blind but develop quite quickly, being able to run and climb at about 4 or 5 weeks old. At 3 months they are weaned and start to eat meat caught by the mother, but stay with her until they are subadult. Leopards are sexually mature at around 3 years of age.

In conflict with humans

Recent decades have seen the leopard's natural habitat, particularly the forests of Southeast Asia, being destroyed on a massive scale by land clearance. In addition, because of attacks on humans, domestic animals and livestock, the leopard has been mercilessly hunted in those parts of its range where it comes into contact with humans. In some places leopards actually confer a direct benefit to farmers because they prey on antelopes, bushpigs and baboons that ravage crops. Despite this, poisoning leopards is a growing problem in some parts of Africa.

The fur trade

Even more intensive hunting has taken place for the fur trade, and this has further reduced leopard numbers. In the 1960s and 1970s it was estimated that some 60,000 leopard skins were being poached each year, often sold by trappers for as little as $1–2 per skin. These were taken to supply the fashion industry, a fur coat selling for more than $60,000 on the black market. A full-sized leopardskin coat requires five to seven skins. Most of the skins were smuggled out through Somalia and Ethiopia because elsewhere leopards were protected by law.

Though the trade in leopard skins is more strictly controlled by governments and international organizations today, there continues to be a thriving black market. Poachers catch the big cats using nooses, springjaw traps, poison and other cruel methods. Several subspecies of leopards are now classified as endangered or critically endangered, including the Barbary or North African, Sinai, South Arabian, North Persian, Anatolian, Sri Lankan, Amur and North Chinese leopards.

Leopards hunt mainly small to medium-sized grazing animals such as deer, gazelles and wildebeest, but also take monkeys, rodents, warthogs (as above), bushpigs and hyraxes.

One well known leopard of Lake Kariba in East Africa seemed to eat nothing but one fish species, in the genus *Tilapia*. It would lie at the water's edge until the fish came to the surface and then catch them with its paw. Another leopard is said to have developed a preference for frogs.

This tendency to specialize may explain the rare instances of leopards that have taken to killing human beings. Once they have killed a human, leopards may develop a taste for the flesh. It might be the same with leopards that kill livestock, because there are known cases of leopards living near farms that never molest the cattle. Some leopards seem to be unusually partial to dogs as food, especially in the outer suburbs of cities such as Nairobi, the capital of Kenya. There are even records of leopards eating other leopards, usually cubs.

LEOPARD FROG

ONE OF THE MOST widespread frogs in North America, the leopard frog has several common names, including grass frog, meadow frog, shad frog and herring hopper. Despite considerable variation in color and markings, some authorities classify the leopard frog as a single species, *Rana pipiens*. However, many scientists argue that the variation in appearance indicates that the frog should be divided into two species. Consequently *R. pipiens* is often referred to as the northern leopard frog and *R. utricularia* is regarded as the southern leopard frog. This article concentrates on the northern leopard frog, although the two forms have a similar lifestyle and habits.

Leopard frogs have slender bodies and pointed heads and measure 2–5 inches (5–13 cm). overall. The color varies, though most commonly they are brown or greenish brown with rows of green, brown or black spots, each ringed with a lighter color. The legs are sometimes bright green and the belly may be white. Leopard frogs are found from Labrador and Mackenzie in southern Canada south through the United States and Mexico to Nicaragua. However, they are not so common along the Pacific coast as they are in the central and eastern parts of the United States.

Returns to water

The leopard frog usually lives near streams, lakes and marshes, and sometimes even along irrigation streams running through deserts. It is found both in mountains and on plains, its range apparently being limited only by its ability to reach water. Adult frogs may wander 1 mile (1.6 km) or more from open water, but the younger ones tend to stay near the banks. However, although adults may stray into drier areas, they return to open water for the breeding season.

In wet environments leopard frogs live in crevices or holes to which they return year after year, while in drier areas they clear away patches of leaf litter to create very shallow depressions in which to rest. In winter leopard frogs hibernate underwater, in mud or beneath stones.

Varied diet

The food of the adult leopard frog includes leeches, snails, spiders and many kinds of insects, both terrestrial and aquatic. These include crickets, grasshoppers, houseflies, beetles, backswimmers and caddis flies, as well as bees and wasps. Leopard frogs also take larger prey, such as tadpoles and small fish, frogs and snakes. They have also been known to catch diminutive birds, such as the ruby-throated hummingbird, *Archilochus colubris*. At the tadpole and froglet stage, leopard frogs are mostly herbivorous, feeding on algae and other plant matter.

A variety of calls

As soon as leopard frogs come out of hibernation they breed, the mating season taking place between mid-March and June. The male has three distinct calls or songs. The main song is a long, low, grunting note followed by several short notes and does not carry very far. Its purpose is to attract other males and females to the breeding pools.

The males indiscriminately grasp any other frog, recognizing them as male or female by touch—the females are swollen with eggs—and by voice. If a male is clasped, it utters a warning call to deter the clasping male. The male also calls to tell the female that he intends to grasp her, and in so doing prepares her for mating. If the female is not ready for mating, she warns the male off by grunting. During mating, the male

Leopard frogs vary in color but always have plenty of large spots. These help to hide the frogs by breaking up their outline.

Leopard frogs are common in many of North America's freshwater wetlands. They venture into drier areas outside the spring breeding season.

clasps the female around the shoulders using his specialized thumbs and executes a series of backward shuffles, swimming backward for a short distance then resting. The female swims with the male attached to her back during mating. The release of her eggs stimulates the male to release sperm, which fertilizes them.

About 3,000 to 6,000 eggs are laid in spherical masses of up to 5 inches (13 cm) in diameter and are fertilized immediately. The egg masses are usually attached to water plants, though they may also float free in the water and often sink to pond bottoms. Depending on the water temperature, the eggs hatch in 2–3 weeks, releasing tadpoles that are about ⅖ inches (1 cm) long. These change into froglets when they are about 1 inch (2.5 cm) long, after 8–11 weeks' growth. The small frogs live in marshes and begin breeding when they are 1–4 years old, although in warmer places they may start breeding after only 1 year. One leopard frog lived in the London Zoo for 9 years.

Attracted to blue light

Leopard frogs can leap distances of up to 8 feet (2.5 m), which is 15 times their body length. By way of comparison, bullfrogs are able to jump nine times their body length. Most leopard frogs live near enough to water for them to be able to reach the safety of the mud with a few leaps.

Experiments have shown that leopard frogs have a simple mechanism that guides them toward water. Their eyes are highly sensitive to blue light. If placed in a box that contains two windows behind which different colored screens are placed, a leopard frog instinctively jumps toward a blue screen more often than it jumps toward any other colored screen. By contrast, the frogs very rarely jumped toward green.

NORTHERN LEOPARD FROG

CLASS **Amphibia**

ORDER **Anura**

FAMILY **Ranidae**

GENUS AND SPECIES **Rana pipiens**

ALTERNATIVE NAMES
Meadow frog, grass frog, shad frog, herring hopper

LENGTH
2–5 in. (5–13 cm)

DISTINCTIVE FEATURES
Large body; 2 ridges of pale skin extend from behind eyes to rear of body; mainly brown or greenish brown, covered with large green, black or brown spots; bold white or yellow stripe on upper jaw

DIET
Mainly insects, leeches, snails and spiders; also small vertebrates, including small fish, snakes, frogs and (rarely) birds

BREEDING
Age at first breeding: 1–4 years; breeding season: mid-March–June; number of eggs: 3,000 to 6,000; hatching period: 2–3 weeks; larval period: about 90 days (depending on temperature)

LIFE SPAN
Up to 9 years in captivity

HABITAT
Still waters of ponds, ditches, canals, marshes, bogs and swamps; also flooded areas during winter and damp fields and meadows during summer

DISTRIBUTION
Much of North America, excluding southeastern U.S. and Pacific coastal regions; south to Texas

STATUS
Common

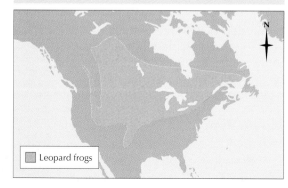

Leopard frogs

LEOPARD SEAL

THE LEOPARD SEAL HAS a reputation for being a voracious predator that is also liable to attack human beings. However, although it is certainly a powerful and extremely efficient hunter, these beliefs are mainly unjustified. The leopard seal is usually solitary and is found all around the Antarctic, often at the edges of the pack ice. It is the only seal in which the adult female is always larger than the male. The male grows to a maximum length of about 10 feet (3 m) and weighs up to 600 pounds (270 kg). The female, on the other hand, may grow as long as 11 feet (3.4 m) and weighs considerably more, perhaps up to 1,300 pounds (600 kg).

Apart from size, the two sexes look much alike. Both are black to gray on the back and paler gray to silver below. The leopard seal has a variable number of darker gray spots, especially on the throat and undersides. It is sometimes described as being almost reptilian in appearance, its body being long, thin and flexible. At first sight its head seems a few sizes too large. The mouth, which can be opened very wide, adds to the ferocious appearance. It has long, pointed teeth, those in the cheeks having three distinct cusps.

Leopard seals live on the outer edges of the Antarctic pack ice. The subantarctic islands of South Georgia, Heard and Kerguelen are also well populated, especially in the winter months. The seals are occasionally found off the southern parts of South America and South Africa, and not infrequently off southern Australia and New Zealand. The leopard seal population has been estimated by observations from polar research vessels to be around 400,000 animals.

Hunting among the ice floes

Leopard seals usually live out to sea among the ice floes and are rarely seen along shores except when they prey on penguins. When they do come onto land or ice floes, it is usually at night. Rarely are more than three or four seen together at one time. Leopard seals move over land or ice with the foreflippers pressed to the side of the body and clear of the ground, and heave themselves forward a little like caterpillars, looping forward on the chest and rear end, alternately doubling up and stretching out.

Powerful predator

The diet of the leopard seal includes crustaceans such as krill, fish, squid and penguins. It will also sometimes take other seal species. The leopard seal has a distensible trachea (windpipe), which collapses to allow the animal to swallow very large chunks of food. It has been known to swallow penguins whole.

Around islands such as South Georgia and the South Sandwich Islands, where there are very large penguin rookeries (colonies), there are also substantial numbers of leopard seals. They lie in wait just offshore and attack the birds as

The leopard seal is one of the top predators in the Southern Ocean. It spends most of its time hunting out to sea, but after feeding it occasionally hauls onto land or ice floes to rest.

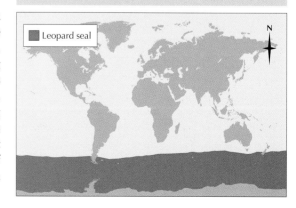 part — table sidebar:

LEOPARD SEAL

CLASS **Mammalia**

ORDER **Pinnipedia**

FAMILY **Phocidae**

GENUS AND SPECIES *Hydrurga leptonyx*

WEIGHT
Male: 440–600 lb. (200–270 kg).
Female: up to 1,300 lb. (600 kg).

LENGTH
Head to tail: 8–11 ft. (2.4–3.4 m); female larger than male

DISTINCTIVE FEATURES
Long, thin body; massive head with long, pointed teeth; black to dark gray above, paler gray to silver below; gray spots on throat and underside

DIET
Crustaceans such as krill; also squid, fish, penguins and other seal species

BREEDING
Age at first breeding: 3–4 years; breeding season: November–February; number of young: 1; gestation period: about 330 days; breeding interval: 1–2 years

LIFE SPAN
Up to 30 years

HABITAT
Cold subantarctic waters; usually among pack ice and seldom comes ashore

DISTRIBUTION
From outer edges of Antarctic pack ice northward to subantarctic islands of South Georgia, Heard and Kerguelen; occasionally as far as southern South America, South Africa, Australia and New Zealand

STATUS
Locally common; estimated population: approximately 400,000

Unlike most other seals, the heavyweight leopard seal is solitary by nature.

they set out for the feeding grounds or as they come ashore after feeding at sea. The leopard seals chase the penguins in the water and grab them from underneath. A penguin can outmaneuver a seal, and many escape to the shore. Only very rarely have leopard seals been seen chasing penguins on land or ice.

It was once thought that penguins were the leopard seal's main food, but it has since been shown that a relatively small number of leopard seals habitually lie in wait near penguin rookeries. Many more fish out at sea. The reason for the mistaken view of this species' feeding ecology is that in the past the leopard seals most often seen were those lingering near the shore.

Icy nursery

Leopard seals breed on the pack ice between November and February. The single pup is born the following year between September and January after a gestation period of about 330 days. It is some 5 feet (1.5 m) long at birth, and weighs about 60 pounds (27 kg). The mother suckles her pup for 2 months before mating again.

Curious, not ferocious

In the few instances in which people have been attacked by a leopard seal it is usually because they have provoked the animal first. Such cases, together with many stories of humans being chased, have helped bolster the seal's ferocious reputation. Leopard seals follow small boats, leaping out of the water and even climbing aboard. They often swim around divers. This is most likely curiosity on the part of the seals, but people seldom put this to the test. The sight of the large head and rows of teeth make discretion the better part of valor.

LIMPET

Limpets are various kinds of snails that, while not necessarily closely related, have two features in common. They cling tightly to rocks or other surfaces, and they have a shell that is more or less tent-shaped. The best known is the common limpet, *Patella vulgata*, also known as the European limpet, flitter, flither or papshell. Related to it is the blue-rayed limpet or peacock's feathers, *Patina pellucida*, which is found on some of the large brown seaweeds at low tide, and the tortoiseshell limpet, *Acmaea tessulata*. The keyhole limpet, *Fissurella costaria*, is named for the hole in the top of its conical shell, while the slit limpet, *Emarginula elongata*, is named for the slitlike opening at the front.

The limpet form just described has also evolved independently in certain relatives of the pond snails. This repeated emergence of the limpet form in the evolution of snails is due to the advantage it gives in withstanding the action of fast-moving or turbulent water. It is particularly advantageous on rocky, wave-battered shores, and it is on these that the common limpet, the main subject of this article, is so successful and abundant.

Strength of the shell design

Under the protection of its ribbed, conical shell, which may reach nearly 3 inches (7.5 cm) in length, the common limpet has a grayish green oval foot with a large flat adhesive surface. At the front is the head with its big earlike tentacles, each bearing an eye near its base. Lining the shell around its margin is the thin layer of tissue that secretes it, and all around this skirt and lying in the space between it and the foot are many small ciliated gills and short tentacles.

The common limpet may occur in great numbers on rocky shores, from low water even to above the highest tide levels, provided the rocks are well splashed and shaded. Shells high up on the shore tend to be taller and thicker, especially near the apex, than those found lower down or in rock pools. As well as being resistant to wide temperature fluctuations, limpets can flourish where the sea is greatly diluted with fresh water. However, they are rarely found far up river estuaries.

Hard to dislodge

Limpets are extremely difficult to dislodge unless taken by surprise, and one scientist found that a weight of 28 pounds (13 kg) could be supported when attached to the shell. As well as giving resistance to wave action, the ability to cling serves to protect limpets from predators. They are, however, a favorite food of oystercatchers (*Haematopus* spp.), and rats may take them in large numbers, dislodging them by a sudden movement of the jaws. In rare instances the tables are turned and a rat becomes trapped with its lip under the shell, and an oystercatcher may be caught by its toes. The limpets' defensive strategy is known as stamping behavior.

Limpets often live together in small groups, and their grazing habits play an important part in controlling the growth of seaweed.

A common limpet crawling over glass, showing the muscular foot and head and marginal tentacles.

Carving a niche

A limpet can seal a little water in its gills and so avoid drying up when the tide is out, and in this it is helped by the close fit of shell to rock. This comes about in the first place from the choice of a suitable resting place, but it is improved both by the growth of the shell to fit the rock and by the abrasion of the rock to fit the shell. Each limpet has a definite home to which it returns after each feeding excursion. Often the rock becomes so worn the limpet comes to lie in a shallow scar of just the right size and shape.

Limpets leave their home when the tide is in and the water not too rough, as well as when the tide is out at night or if they are sheltered by seaweed. They feed by rasping at the small green algae on rocks, moving around with head and tentacles protruding and swinging from side to side. They prevent large seaweeds from covering areas of rock by eating them when they are small. On the return journey, each limpet tends to retrace its outward track. It seems to have a kind of chemical sense for feeling its way back.

Breeding

The common limpet breeds during the colder months, shedding its eggs freely into the sea. The tiny larvae hatch out about 24 hours after fertilization. A day or so later, they start to develop little shells. These settle and grow to about 1 inch (2.5 cm) long in a year. Most limpets start life as males and remain this way until they are an inch long. As they get older, the proportion of females increases, because some limpets reverse their sex. This process goes much further in the slipper limpet or slipper shell, *Crepidula fornicata*, of North American seas. It forms chains of up to nine, one on top of another. The bottom ones are females and the upper, younger ones are males.

LIMPETS

PHYLUM **Mollusca**

CLASS **Gastropoda**

SUBCLASS **Prosobranchia**

ORDER **Archaeogastropoda**

FAMILY **Typical limpets, Patellidae; keyhole limpets, Fissurellidae; slipper limpets, Calyptraeidae; others**

GENUS AND SPECIES **Many, including common limpet, *Patella vulgata* (detailed below); tortoiseshell limpet, *Acmaea tessulata*; keyhole limpet, *Fissurella costaria*; and slipper limpet, *Crepidula fornicata***

ALTERNATIVE NAMES
European limpet; flitter; flither; papshell

LENGTH
Up to 2⅓–2¾ in. (6–7 cm)

DISTINCTIVE FEATURES
Conical, tent-shaped shell; numerous ribs radiate outward from apex (topmost point) of shell; usually pale or whitish in color

DIET
Microscopic algae growing on rocks; small new growths of various seaweed species

BREEDING
Sexes separate. Breeding season: April–September; larval period: varies according to environment.

LIFE SPAN
Usually up to 15 years

HABITAT
Rocky shores, mainly in intertidal zone; also in lower reaches of river estuaries

DISTRIBUTION
Eastern coasts of North Atlantic, from Iceland and Scandinavia south to Portugal

STATUS
Very common

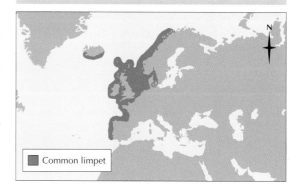

Common limpet

LIMPKIN

THE LIMPKIN IS THE sole member of the New World family Aramidae, and appears to be a link between the rails and the cranes. A limpkin is ibislike, with long legs, a fairly long neck and a long, stout and slightly curved bill. It typically stands about 2–2⅓ feet (60–70 cm) high. The plumage is olive brown, finely streaked with white on the head and upperparts. Juveniles are paler than adults.

The single species of limpkin lives in Florida and southern Georgia, and then ranges from Mexico south through Central America and South America east of the Andes to northern Argentina. It also occurs in the Caribbean.

Eerie marsh bird
The limpkin lives in freshwater marshes and swamps and sometimes in damp forests. It was once common in Georgia and Florida but was good eating and easy to shoot. Its survival is, however, now ensured in sanctuaries such as the Everglades National Park in Florida and Okefenokee Swamp in Georgia, provided that these wetlands are not drained.

The name limpkin is derived from the bird's strange, limping gait as it treads gingerly across the matted swamp vegetation, lifting its long toes high and twitching its tail. Limpkins swim if necessary, floating high like coots or gallinules. They rarely fly, but when they do, they take off and land vertically and fly weakly with slow steady beats. At night limpkins roost in trees, where they can run among the branches with surprising agility.

Other names of the limpkin are crying bird and wailing bird, and in Mexico it is traditionally called the "mad widow." These names refer to the eerie, wailing calls limpkins make, mainly at night. The calls are of three syllables and have been described as shrieks, wails and piercing cries, or as having "a quality of unutterable sadness." According to folklore, the cries were made by little boys lost forever in the swamps.

Diet of snails
Limpkins feed at wading depth in the shallow waters of swamps and marshes. Their main food is the same large freshwater snail, the apple snail (*Pomacea* sp.), that is the exclusive diet of the Everglade snail kite, *Rostrhamus sociabilis*. The kite can feed on the snails only when they come out to feed in the early morning and late afternoon, but the limpkin can probe for them in the mud with its long bill. The diet of snails is supplemented with worms, crayfish, insects and occasional amphibians and reptiles, but these items are of relatively little importance.

Hunting technique
The feeding grounds of limpkins are very conspicuous as they become littered with small piles of snail shells. Limpkins are surprisingly tame except when hunted, and it is not too difficult to watch limpkins feeding from a boat. They feed singly or in small flocks. As the snails are caught, they are held by the flange of the shell and carried to land or shallower water. They are lodged carefully in a crevice or in the fork of a branch, with the opening upward. The limpkin then waits patiently for the snail to relax and spears it, passing the upper half of its bill through the body between the shell and the horny operculum (door). A jerk of the head detaches both operculum and shell, but instead of swallowing the body, the limpkin waits a couple of minutes before consuming it.

Limpkins wade in the shallows of marshes and swamps looking for a particular type of snail. They rarely eat anything else.

Secluded places overgrown with lush vegetation make ideal hunting grounds for the limpkin. It rarely flies, preferring to walk.

LIMPKIN

CLASS	**Aves**
ORDER	**Gruiformes**
FAMILY	**Aramidae**
GENUS AND SPECIES	***Aramus guarauna***

ALTERNATIVE NAMES
Crying bird; wailing bird

LENGTH
Head to tail: 23–28 in. (57–70 cm)

DISTINCTIVE FEATURES
Resembles an ibis or large rail; long, thick, downcurved bill; long neck; long legs; feet are not webbed but have elongated toes and claws; chocolate brown plumage with numerous white flecks and streaks on head and upperparts

DIET
Mainly freshwater snails; also frogs, worms, crayfish, small lizards and aquatic insects

BREEDING
Age at first breeding: 2 years; breeding season: eggs laid February–April (U.S.); number of eggs: usually 4 to 6; incubation period: not known; fledging period: not known; breeding interval: probably 1 year

LIFE SPAN
Up to 12 years

HABITAT
Shallow margins of freshwater marshes, pools and swamps; also damp forests

DISTRIBUTION
Southeastern U.S. south through Caribbean and Central America to northern Argentina; absent west of Andes Mountains

STATUS
Locally common in most of range; generally rare in U.S. due to hunting and habitat loss, although recovering in places

Precocious chicks

Virtually nothing is known of the courtship of the limpkin, but the nest is built of reeds, grasses and other plant material. It is bulky but flimsy and is lodged in tangled vines, grass clumps, a bush or a tree, sometimes as much as 17 feet (5.2 m) up and usually near or over water. The clutch consists of up to eight, but usually four to six, pale green eggs spotted with brown. In the United States they are laid between mid-February and April and are incubated by both sexes, but the incubation period is unknown.

The brownish black chicks can swim and run almost immediately and after a day or so leave the nest and hide nearby. They are guarded by the parents for an unknown length of time, but it is known that the parents feed them after they have begun to fly. The chicks approach the parents from behind and reach forward between their legs to take the proffered snails.

Changing fortunes

With such specialized dietary and habitat requirements, both the Everglade snail kite and limpkin have precarious existences. The Everglade snail kite is now endangered in Florida because of the draining of the swamps, although it is often locally common in the rest of its range in the tropical Americas. The limpkin is more plentiful, however, and its numbers are recovering now that it is protected. It is even abundant in places where food is in good supply and human disturbance is minimal.

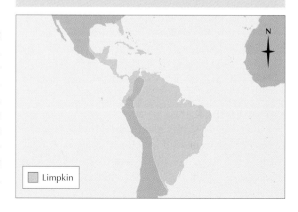

Limpkin

LING

THE LING ARE THREE species of large, elongated fish in the family Lotidae. They are closely related to the cod family, Gadidae, which contains 150 species, including important commercial species such as the cod, coalfish or pollack, haddock, hake and whiting.

Ling have a somewhat eel-shaped body and there are two dorsal fins, the one in front being short, while the one behind is very long. The single anal fin is long. Ling have the characteristic barbel on the chin of all members of the Lotidae family. They are commercially valuable and are also popular as sport fish. Ling live at depths of between 330 and 1,650 feet (100–500 m), depending on species. Here they can be fished by trawl, although they are more often taken by longline. Small ling are caught offshore, in shallow waters, by anglers.

North Atlantic fish

All three species are found in the northeastern Atlantic. The common ling, *Molva molva*, ranges from the Barents Sea and Iceland to as far south as Morocco. It is also occasionally found in the northwestern Mediterranean, off southern Greenland and Canada. It lives mainly on rocky bottoms in fairly deep water and is a dull brownish green color, mottled on the back and lighter below. Its fins are spotted, dark on the first dorsal fin, but with paler spots on the other fins.

The blue or lesser ling, *M. dypterygia*, is found on muddy bottoms at depths of 1,150–1,650 feet (350–500 m). It lives in the southwestern Barents Sea, sometimes north to Spitzbergen, southeastern Greenland and the southern coast of Iceland. It is also found around the British Isles and ranges south to Morocco and into the Mediterranean. In the northwest Atlantic it is found off Newfoundland. It is similar in color to the common ling, but is plain rather than mottled and has dusky patches on the second dorsal and anal fins.

The Spanish or Mediterranean ling, *M. macrophthalma*, occurs in moderately shallow waters to the south of the other two species. It is sometimes caught in the eastern Atlantic as far north as the south of Ireland, and is also found in the western Mediterranean. While the common ling might reach up to 6½ feet (2 m) in length, both the blue and Spanish ling are somewhat smaller. The blue ling grows to a maximum of 5 feet (1.5 m), while the Spanish ling reaches only 3 feet (0.9 m).

A fish similar to the ling is the burbot, *Lota lota*, or eel pout. It is the only freshwater member of the cod family and is found in large rivers and deep lakes around the Arctic region. It is shaped like the ling and is their nearest relative.

Squeeze between rocks

Common ling often lie still for hours with as much as possible of their bodies in contact with hard objects. They are often found around rocks and sheltering in the cracks between them.

The common ling, a long, eel-shaped fish, is most often found in rock crevices in fairly deep water.

Ling are renowned for their fecundity, the female spawning up to 60 million eggs.

Ling press themselves into crevices and even curl around stones, following the curves as closely as possible. This may be the reason why ling are common in the neighborhood of submerged wrecks, for they must find excellent places among the old girders on which to lie.

The common ling feeds mainly on smaller fish, especially young cod, herring, mackerel and flatfish, but it also takes lobsters, cephalopods and starfish. The blue ling tends to feed more on crustaceans, along with a few fish such as flatfish, gobies and rocklings.

Mother to millions

Ling are renowned for their fecundity, each female laying a remarkable number of eggs. It is thought that a single female may contain more than 60 million eggs. Spawning of the common ling is from March to June, or later, sometimes into August, in more northern latitudes. Spawning takes place mainly at depths of 600 feet (180 m). The eggs are about 1 millimeter in diameter and each contains a colorless or pale green oil globule carrying black pigment. The oil globule causes the eggs to float to the surface but the young go down into deeper water soon after hatching. This takes place when the eggs have swollen to about 3 millimeters in diameter.

Young common ling, when they are living at 120–300 feet (35–90 m) in depth, grow very long pelvic fins. As time passes and the fish grow, these fins are lost. Finally the young ling reach the seabed when about 3 inches (7.5 cm) long.

COMMON LING

CLASS	**Osteichthyes**
ORDER	**Gadiformes**
FAMILY	**Lotidae**
GENUS AND SPECIES	*Molva molva*

WEIGHT
Up to 66 lb. (30 kg), usually much less

LENGTH
Up to 6½ ft. (2 m)

DISTINCTIVE FEATURES
Rather codlike but with longer body; single barbel on chin; 2 dorsal fins, first short and second very long; dull brownish green overall; mottled above, lighter below

DIET
Smaller fish such as young cod, herring, mackerel and flatfish; also lobsters, cephalopods and starfish

BREEDING
Age at first breeding: not known; breeding season: March–August; number of eggs: up to 60 million; hatching period: eggs hatch when 3 mm in diameter; breeding interval: 1 year

LIFE SPAN
Up to 14 years

HABITAT
Rocky seabeds in fairly deep waters; usually at depths of 330–1,300 ft. (100–400 m)

DISTRIBUTION
North Atlantic, from Barents Sea and Iceland south to Morocco; occasionally off southern Greenland and Canada and into northwestern Mediterranean

STATUS
Relatively common in parts of range; not threatened despite commercial fishing

Common ling

LINNET

LINNETS CAN BE ALL around and yet remain unseen, or at least undetected. This is partly because of their shyness and partly the result of their predominantly brown plumage. Linnets are finches and belong to the subfamily Carduelinae, or typical finches.

Different plumages

Smaller than the house sparrow, *Passer domesticus*, linnets are 5–5½ inches (13–14 cm) long. In winter the male has a brown back and is fawn shading to almost white on the underside, with a grayish brown head mottled or streaked with darker brown. The flight feathers and tail have white edges. This is his winter plumage and then the bill is horn-colored. By the spring he boasts a crimson forehead and breast, the bill then being lead-colored. The crimson patches appear as the gray ends of the feathers are worn away.

The female is slightly smaller than the male. She also lacks the crimson areas of the breeding male, her plumage is generally duller and grayer and the brown streaks on her head and breast are more obvious.

The name linnet is derived from the Old French and is based on *lin*, meaning flax, presumably because the bird eats the seeds of flax. It has also been applied to several other related finches. The greenfinch, *Carduelis chloris*, was sometimes traditionally called the green linnet. The twite, *C. flavirostris*, was once called the mountain linnet, and the North American pine siskin, *Spinus pinus*, is occasionally known as the pine linnet.

Birds of open country

Linnets range over much of Europe, except for the extreme north, and are also found in northwestern Africa, southwestern Asia and part of Central Asia, from sea level up to the tree line. They live in most open habitats where there are scattered bushes or trees, including on moors, heaths and hillsides, at the edges of copses and plantations and in hedgerows, gardens and cultivated land with plenty of rank vegetation, such as gorse and bramble. In winter, when linnets come together in flocks, they move across arable land and stubble fields, and often occur along the coast on salt marshes and rough grassland.

Sweet song

Male linnets are noted for their song. One 17th-century writer put the linnet second only to the common nightingale, *Luscinia megarhynchos*, which is one of the finest singers of Europe and Central Asia. Linnets sing while perched high on a bush, on the ground or on the wing. Several males may sometimes be seen in the same bush singing in a chorus that does not carry far but is pleasing and musical. Some of the high notes are flutelike, whereas some of the low notes are rather harplike.

Formerly linnets were valued as cage birds and there was a belief that a linnet crossed with a canary, *Serinus canaria*, produced an even better song than either of the parent species. In fact, the song of a linnet is mainly learned, not inborn, so it could not be affected much by inheritance.

Seed eaters

Typically seed eaters taking small seeds of garden weeds, but more especially flax, hemp and brassicas such as turnips and cabbage, linnets will also take berries and oats in winter. They eat some insects, and the young are fed on caterpillars, fly larvae, small beetles and spiders.

In breeding plumage male linnets have crimson patches on the breast and forehead. They often perform their sweet song from the top of bushes.

LINNET

CLASS	**Aves**
ORDER	**Passeriformes**
FAMILY	**Fringillidae**
SUBFAMILY	**Carduelinae**
GENUS AND SPECIES	*Carduelis cannabina*

WEIGHT
½–⅔ oz. (16–18 g)

LENGTH
**Head to tail: 5–5½ in. (13–14 cm);
wingspan: 8¼–10 in. (21–25 cm)**

DISTINCTIVE FEATURES
**Breeding male: gray head; warm brown
upperparts; crimson forehead and breast
patch. Female and nonbreeding male:
grayish brown above and pale below with
many brown streaks.**

DIET
**Seeds of grasses and weeds, especially flax;
some buds and invertebrates; rarely berries**

BREEDING
**Age at first breeding: 1 year; breeding
season: April–September; number of eggs:
4 to 6; incubation period: 10–12 days;
fledging period: 11–12 days; breeding
interval: up to 3 broods per year**

LIFE SPAN
Probably up to 8 years

HABITAT
**Breeding: open areas with scrub and
scattered trees, especially hillsides,
moors, heaths and edges of cultivation.
Winter: mainly fields, coasts and plains.**

DISTRIBUTION
**Europe and North Africa east to Central
Asia and the northern Middle East**

STATUS
Common; declining in northwestern Europe

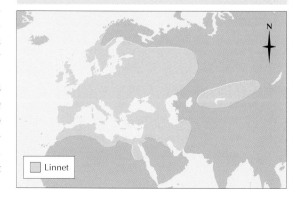

☐ Linnet

Linnets are specialist seed eaters and spend a lot of their time foraging on the ground.

Hidden nests

The breeding season begins in April, and eggs may be laid any time up to August. The well concealed nest is built of grasses and moss, sometimes also with fine twigs, and is then lined with wool, hair and down or feathers. It is sited in a bush or shrub, not more than a few feet from the ground.

The usual clutch size is four to six eggs, and up to three clutches are laid per year. The eggs are pale bluish with spots, occasionally streaks, of purplish red. These are incubated for 10–12 days, mainly by the hen (female), the cock (male) relieving her for only short periods so that she can exercise and drink. The young are fed by both parents for 11–12 days.

Close neighbors

Linnets tend to nest in small groups, with several nests located only a few feet from each other. Pairs defend the area around their own nest but, unlike many other finches and songbirds in general, they do not defend a breeding territory in the strict sense of the word.

Normally linnets nest well away from houses but they will occasionally build in a bush beside a porch or near a window. Such nests help to show how shy linnets can be because even occupants of the house, peering through a window, have difficulty in watching the birds on the nest or flying to and from it.

LINSANG

The banded linsang (above) is slightly larger than the spotted species and ranges farther south into the island of Java in Indonesia.

THE LINSANGS ARE PROBABLY the most slender members of the family Viverridae, which comprises 35 species and includes the civets, palm civets and genets. There are two species: the spotted and banded linsangs.

The banded linsang, *Prionodon linsang*, is about 2½ feet (75 cm) in total length, of which nearly half is tail. It weighs 1¼–1¾ pounds (600–800 g). Its body is long and slender, its legs are short and it has a pointed snout. The fur is short, dark and thick, with broad, irregular whitish to brownish gray bands across the back and neck. Its thick tail is ringed white and blackish brown. The slightly smaller spotted linsang, *P. pardicolor*, is similar in most respects but is dark with longitudinal rows of pale brown or orange buff spots on the back and flanks.

Both species are found in the tropical forests and jungles of Southeast Asia. The banded linsang ranges from Thailand south to Java and Borneo. The spotted linsang is found farther north, from Nepal eastward through Assam, India, to southern China.

Nocturnal hunters

Linsangs are solitary, nocturnal prowlers, resting most of the day in tree hollows. They are agile climbers but spend most time on the ground. They hunt small mammals, birds, frogs, lizards and insects. Fish and eggs may also be taken.

The linsangs are rare or threatened in most parts of their range and little is known of their reproduction. There is no distinct breeding season and two to four young are usually born. The baby linsangs are raised in a nest of sticks and leaves in a hollow tree or in a hole in the ground.

LINSANGS

CLASS **Mammalia**

ORDER **Carnivora**

FAMILY **Viverridae**

GENUS AND SPECIES **Banded linsang, *Prionodon linsang*; spotted linsang, *P. pardicolor***

ALTERNATIVE NAME
Oriental linsang (both species)

WEIGHT
1¼–1¾ lb. (600–800 g)

LENGTH
Head and body: 1–1½ ft. (30–45 cm); tail: 1–1⅓ ft. (30–40 cm)

DISTINCTIVE FEATURES
Resembles a very long-bodied, slender cat; thick, dark coat with pale stripes (striped linsang) or spots (spotted linsang); prominent ears; large eyes; long, thick tail with alternate light and dark rings

DIET
Small mammals, birds, lizards, frogs, eggs and invertebrates; sometimes fish

BREEDING
Age at first breeding: not known; breeding season: all year; number of young: 2 to 4; gestation period: not known

LIFE SPAN
Up to 11 years in captivity

HABITAT
Tropical rain forests and jungle

DISTRIBUTION
Banded linsang: Thailand south to Peninsular Malaysia, Borneo and Indonesia (including Sumatra and Java). Spotted linsang: eastern Nepal east through Assam, India, to southern China.

STATUS
Rare in most parts of range

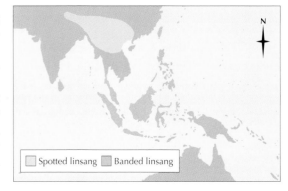

☐ Spotted linsang ☐ Banded linsang

LION

Lionesses (left, above) do most of the hunting for each pride, or group of lions. A pride consists of several generations of related lionesses and one or more adult males.

THE LION USED TO have an extremely wide geographical distribution. It was once common throughout southern Europe and southern Asia eastward to northern and central India and was found across the whole of Africa. At one time it also ranged throughout North America, but it is thought to have disappeared from this region some 10,000 years ago. The last wild lion died in Europe more recently, probably between 80 and 100 C.E. By 1884 the only lions left in India were restricted to the Gir Forest in Gujarat, where only a dozen or so remained. It is thought that the lion probably became extinct elsewhere in southern Asia, for example, in Iran and Iraq, soon after that date.

Changing fortunes

Since the beginning of the 20th century the Gir lions, the only remaining Asiatic lions, have been protected. Recently they were estimated to number about 250. In Africa, meanwhile, the lion has been wiped out in the northern part of the continent and its numbers are now estimated to stand at around 50,000 in sub-Saharan Africa. Although protected in areas such as the Kruger National Park in South Africa and Tanzania's Serengeti, populations are increasingly isolated.

Unmistakable big cats

Lions are unmistakable. The male, or lion, may reach more than 8 feet (2.5 m) in length, with a 3-foot (0.9-m) tail. It stands about 4 feet (1.2 m) at the shoulder and can weigh up to 550 pounds (250 kg). The female, or lioness, is smaller, up to 6 feet (1.8 m) in length and standing 3–3½ feet (0.9–1.15 m) at the shoulder, with a slightly shorter tail. The lion's coat is short and tawny, orange brown, buff or grayish. There is a dark tuft at the tip of the tail. Lionesses are more generally sandy colored. Dominant males have a pronounced tawny to black mane over the head, neck and shoulders, which may extend to the belly. Maneless lions occur in some districts.

The only truly sociable cats

Lions live in open country and savanna with scrub, spreading trees or reed beds. The only truly sociable member of the cat family, they live

LION

CLASS	**Mammalia**
ORDER	**Carnivora**
FAMILY	**Felidae**
GENUS AND SPECIES	***Panthera leo***

WEIGHT
Male: 330–550 lb. (150–250 kg).
Female: 265–400 lb. (120–180 kg).

LENGTH
Male, head and body: 5½–8 ft. (1.7–2.5 m);
shoulder height: about 4 ft. (1.2 m).
Female, head and body: 4½–6 ft. (1.4–1.8 m);
shoulder height: 3–3½ ft. (0.9–1.15 m).

DISTINCTIVE FEATURES
Very large size; sandy, tawny, buff or
grayish coat with no spots or stripes.
Dominant male: heavyweight body; shaggy
mane on head and neck. Female and young
male: more lithe and muscular; no mane.

DIET
Mainly large grazing animals such as zebra,
impala and wildebeest; also young buffalo,
other antelopes, carrion and fallen fruits

BREEDING
Age at first breeding: 2 years (female),
usually about 5 years (male); breeding
season: all year, but all females of a pride
mate at same time; number of young:
usually 3 or 4; gestation period: 100–120
days; breeding interval: often 2 years

LIFE SPAN
Up to 30 years in captivity

HABITAT
Savanna, scrub, open bush and forest

DISTRIBUTION
Sub-Saharan Africa; Gir Forest, Gujarat, India

STATUS
Africa: vulnerable; split into isolated
populations. India: endangered.

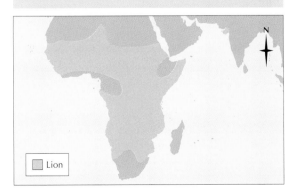

in groups known as prides, often comprising about 15 animals, although there may be up to 30. A pride consists of one or more mature lions and several generations of related lionesses with their cubs. There will also be some young males, which stay with the pride until 2 or 3 years of age. Females may remain with the same pride for life, or may move to another pride on reaching maturity. The adult males are outsiders and will hold the pride only so long as they are able to defend it against other outside males. A pride's well-defined territory will vary in size according to the abundance of prey in the region. Members of a pride often cooperate with one another when hunting, stalking or ambushing prey, and also work together to defend the group.

Lionesses have been heard to communicate with one another by grunting when stalking prey, but the characteristic roar of the male is most often used to proclaim territory. Roaring can be heard at dawn and in the evening before a night's hunting. A lion is capable of speeds of up to 40 miles per hour (65 km/h), but is able to run only short distances before tiring. It can make standing jumps of up to 12 feet (3.6 m) and is

Once widespread throughout southern Asia, the Asiatic lion is now restricted to the Gir National Park in Gujarat, India. The population of 250 animals is strictly protected.

also able to leap distances of some 40 feet (12 m). Lions rarely, if ever, climb trees, but lionesses may jump onto low branches to sun themselves.

Carnivorous diet

Lions are mainly carnivorous, but take fallen fruit at times. They obtain all necessary protein, fat, carbohydrate, mineral salts and vitamins from the flesh and entrails of the animals they kill.

Typically, lions eat the entrails and hindquarters of their victims first, working forward to the head of their prey. It is usually the lionesses that hunt together and make the majority of kills. However, other members of the pride quickly move in and the larger males often take most of the meat. The lionesses next take their place at the carcass. The cubs come last, sometimes getting little or none of the kill.

The larger males often devour the majority of a kill, even when they have not participated in the hunt. The lionesses feed after the males, the cubs having to settle for whatever remains.

In general, lions feed on large grazing animals such as zebras, wildebeest and antelopes. However, almost any animal will be taken, from cane rats to elephants, hippos, giraffes and even ostriches. Lions also scavenge from kills made by hyenas. A survey in the Kruger National Park showed that in order of numbers killed the prey-species were: wildebeest, impala, zebra, waterbuck, kudu, giraffe and buffalo. A later survey showed a preference as follows: waterbuck, wildebeest, kudu, giraffe, sable, tsessebe or topi (a species of large antelope), zebra, buffalo, reedbuck and impala.

When age or injury prevents a lion from catching agile prey, it may turn to porcupines and smaller rodents. In some regions domestic sheep, goats and dogs are taken. Very occasionally a lion will take a human, with man-eating becoming a habit in certain individuals. Once a group of lions at Tsavo in Uganda held up the building of a railway because of their attacks on the laborers.

Hunting methods

A large number of hunts end in failure because lions tire after running only short distances. A lioness will stalk her prey using the brush and scrub for cover. When close, she will run at the animal in a short, rapid dash. When hunting in a group, the lionesses encircle a herd of zebra for example, approaching it from opposite directions. They then close in and attempt to bring one down as the herd panics. The usual method of killing is for the lioness to leap at her prey and break its neck using her front paws. Alternatively, she may seize her victim by the throat with her teeth, or she may throttle it. Another method is to leap at the hindquarters, pulling the prey down.

Synchronized breeding

Lions begin to breed at 2 years of age but reach their prime at 5 years. A male is not usually large or strong enough to take over a pride before this. There is no particular breeding season, but mating and births are synchronous between the females of a particular pride and take place around the same time. Both

males and females are polygamous, although females are restricted to breeding with the one or two adult males of the pride. There is a good deal of roaring before and during mating, and fights between the dominant males and intruding males often take place. The gestation period is 100–120 days and three or four, occasionally up to six, cubs are born in a litter.

The cubs are blind and have a spotted coat at birth. Their eyes open after 6 days. Weaning takes place when the cubs are around 6 or 7 months old, after which the lioness teaches them to hunt. Young lions can usually hunt for themselves at 1 year old. There is a high death rate among cubs, which feed last and may suffer from a dietary deficiency, especially of vitamins. This serves as a check on numbers. Should numbers fall in a district, prey is more easily killed by the remaining animals. Lionesses will then kill for their cubs and in this situation the cubs eat first. This richer diet makes for a higher survival rate among the cubs, thus restoring population numbers.

No natural predators

The lion has no predators as such, apart from human beings, but lions, especially the young and inexperienced, are prone to casualties. A zebra stallion may lash out and kick a lion in the teeth, after which it may be able to hunt only small game. The sable antelope is more than a match for a single lion, and other antelopes have sometimes impaled lions on their horns. A herd of buffalo may also trample a lion or gore it with their horns until it is dead. However, two lions can sometimes overcome one large buffalo. Female giraffes have been known to attack lions when protecting their calves.

Threats to African populations

Despite this lack of natural predation, lion populations in southern Africa are on the decline and are now split into isolated sub-populations. This decline has been mainly caused by habitat loss to agriculture and by persecution by humans protecting their livestock. In addition, fragmentation of populations has led to a loss of genetic variation, with smaller numbers of animals breeding together. This is reducing the viability of existing populations. Outbreaks of canine distemper and related viruses have also reduced the numbers of lions in some regions. Nonetheless, the future of Africa's lions in its national parks at least seems secure because of this species' popularity as a tourist attraction.

Lions hunt for only 2 or 3 hours each day and often rest for several days after gorging themselves on a kill.

LITTLE AUK

Despite its tiny size, the little auk or dovekie is tough enough to live in the cold seas of the northern Arctic.

THE LITTLE AUK OR DOVEKIE is by far the smallest of the Atlantic auks, only 6⅔–7½ inches (17–19 cm) long. Its bill is short, stubby and almost finchlike, and this, combined with its compact body, makes it look very different from other auks. However, in flight or from a distance it may be mistaken for a young guillemot or common murre, *Uria aalge*. The little auk's plumage is black with white underparts and white patches on the wings. A small white patch over the eye, seen only at close quarters, gives it a comical appearance. In the winter the upperpart of the breast, throat and sides of the head become white.

Little auks breed in the Arctic, around the coasts of St. Lawrence Island, western Greenland, Jan Mayen, Bear Island, Svalbard, Spitzbergen, Novaya Zemlya, Franz Josef Land and Grimsey, off northern Iceland. Outside the breeding season they range south into the North Atlantic to the coasts of the British Isles in the east and those of the New England states in the west. Very occasionally little auks come farther south, as far as the Mediterranean and Cuba.

Spectacular breeding colonies

Little auks are perhaps the most abundant birds of the North Atlantic, their only rivals for this title being Brünnich's guillemots or thick-billed murres, *Uria lomvia*. In temperate regions they are seen offshore in large flocks, but these are not half as spectacular as the immense numbers that haunt the colonies on the cliffs around the Arctic seas. The cliffs of Greenland and other Arctic islands may be hundreds or even thousands of feet high, and they are packed with various species of seabirds, including the little auks that make their nests on the narrow ledges or among rocks and scree. Surrounding the cliffs are clouds of birds flying to and from their nests while the air is filled with their twittering. One ornithologist has described the scene as being like a vast opera house, packed with crowds of people in white shirt-fronts and black tails, all whispering comments on each other and rustling their programs.

The little auks are not as obvious as the other seabirds because they nest in crevices, but one colony, at Scoresby Sound on the east coast of Greenland, contains an estimated 5 million little auk nests, while there are uncounted millions in the many other colonies. The total breeding population in the Arctic numbers several tens of millions of pairs.

Wrecks of auks

Occasionally flocks of little auks are blown inland by gales, and many perish. These "wrecks," as they are called, occur when the little auks are caught on a lee shore and are unable to maneuver out to sea against the wind. The stranded birds turn up on ponds and lakes, on roads and fields or in gardens. Those that can land on water may be lucky enough to be able to take off when the weather improves, but most die of starvation and exhaustion or fall prey to cats, dogs, foxes, crows and other predators.

Feeding along the pack ice

Little auks feed mainly on crustaceans, such as amphipods, copepods and northern species of krill. Small fish, worms and other planktonic animals are also caught. Like the related guillemots or murres, little auks chase their prey underwater, swimming with their wings.

The main feeding grounds of the little auks are along the edges of the pack ice, where there are massive concentrations of planktonic animals during the Arctic summer. The importance of these shallow waters is shown by the effect that the general retreat of the pack ice (due to global warming) is having on several little auk populations. Numbers of little auks have fallen at a

LITTLE AUK

CLASS	**Aves**
ORDER	**Charadriiformes**
FAMILY	**Alcidae**
GENUS AND SPECIES	*Alle alle*

ALTERNATIVE NAME
Dovekie (U.S. only)

WEIGHT
5–6⅔ oz. (140–190 g)

LENGTH
**Head to tail: 6⅔–7½ in. (17–19 cm);
wingspan: 16–19 in. (40–48 cm)**

DISTINCTIVE FEATURES
**Small, compact body; stubby bill and steep
forehead give snub-nosed profile; webbed
feet set far back on body. Summer: black
head, chest and underparts; white belly.
Winter: sides of head and chest turn white.**

DIET
**Crustaceans contained in plankton; also
very small fish, worms and mollusks**

BREEDING
**Age at first breeding: not known; breeding
season: eggs laid June–early July; number
of eggs: 1; incubation period: 28–31 days;
fledging period: 23–30 days; breeding
interval: 1 year**

LIFE SPAN
Not known

HABITAT
**Shallow Arctic seas; nests on sea cliffs and
rocky scree slopes**

DISTRIBUTION
**Breeding: northern Arctic seas and islands.
Nonbreeding: moves southward to southern
Arctic and cold temperate waters.**

STATUS
Abundant on main breeding grounds

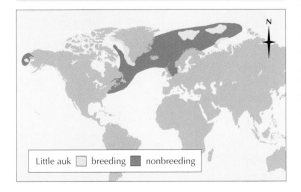

Little auk ☐ breeding ■ nonbreeding

variety of breeding colonies due to the increasing distance between the pack ice and the colony, making it difficult for the little auks to collect enough food for their young.

Crevice nests

Little auks nest in crevices on worn rock faces, among the stones of scree slopes or in the talus, the pile of boulders that accumulates at the base of a cliff. The nests are usually high on the cliffs and may be found on inland cliff faces some distance from the sea. Sometimes the little auks dig out the soil among the stones to make a better hole, and they then make a nest of pebbles about 1 yard (0.9 m) from the entrance.

A single pale blue egg is laid in June or early July. It hatches 4 weeks later after incubation by both parents, which feed the chick on plankton eight or nine times a day, bringing the food to the nest in cheek pouches. The chick leaves the nest when 23–30 days old, making a perilous flight down to the sea on uncertain wings.

Young are vulnerable

In general seabirds that nest on cliffs are safe from most predators. However, the eggs and chicks of little auks are taken by parasitic jaegers, *Stercorarius parasiticus*, and glaucous gulls, *Larus hyperboreus*, which can land and forage among the rocks. The glaucous gulls also kill large numbers of the young little auks as they set off on their first flight. Arctic foxes, *Alopex lagopus*, also hunt on the more accessible scree slopes and cliff ledges, while in parts of Greenland humans take both chicks and adults and store them for use during the winter.

Large breeding colonies have more than 100,000 pairs of little auks. There is a constant stream of parent birds flying to and from their nests.

LIZARDS

LIZARDS ARE REPTILES with scaly skins. They are common in many parts of the world and most species lay eggs. More than 3,000 species have been described, and several dozen new ones are discovered every year. Lizards live in almost every kind of land habitat except for very cold places in the Arctic and Antarctic. They seldom occur above the snow line in high mountains, but can be found up to altitudes of about 15,000 feet (4,600 m) in the Andes and Himalayan mountain ranges. Lizards are particularly abundant in tropical countries, and also in Australia, where at least 500 species are found.

Most lizards are comparatively small: the majority weigh less than 1 pound (460 g) and are less than 18 inches (45 cm) in total length. The smallest species are geckos belonging to the genus *Sphaerodactylus*. Found in the Caribbean, some of these never grow longer than 1 inch (2.5 cm). Komodo dragons, *Varanus komodoensis*, are the world's largest lizards and may grow to up to 10 feet (3 m) in length and can weigh up to 300 pounds (140 kg). They are a species of monitor lizard.

Diverse structure

A typical lizard is elongated, with four legs and a long tail, but there is huge diversity within the group. Some lizards have only very small legs, and in others the legs are entirely absent. Legless lizards look superficially very like snakes. However, they can be distinguished from snakes because they have eyelids and an external ear opening. They also have more than

Lizards are physically diverse, both in size and structure. Two genera have even evolved winglike flaps of skin supported on ribs, enabling them to glide between trees. Pictured is a flying agama lizard of the genus Draco, *which lives in Southeast Asia.*

one row of scales across their undersides and their forked tongues are relatively short and stubby, compared to the long forked tongues of snakes. Internally there are also many differences: most snakes have only one lung, while lizards have two, and snakes have a highly modified skull and jaw structure so that they can swallow large prey.

Most species of lizards feed on invertebrates of various kinds, although a few are herbivorous. The majority are diurnal (active by day), but some are nocturnal, including most of the geckos, family Gekkonidae. Because they are active at night, geckos have large eyes, the pupils contracting to a vertical slit during the daytime. Geckos also have adhesive pads on

CLASSIFICATION

CLASS
Reptilia

ORDER
Squamata

SUBORDER
Sauria

FAMILY
About 17

NUMBER OF SPECIES
Slightly more than 3,000

the undersides of their toes, enabling them to climb with great agility. In the Tropics geckos enter houses, and run upside down across ceilings.

The most highly adapted lizards are the chameleons of the family Chamaeleontidae. They are flattened from side to side, have opposed toes that enable them to cling to small branches, and move only very slowly. Chameleons also have curious eyes that swivel independently and long tongues that can be extruded considerable distances to help them capture their insect prey. They are well known for their ability to change their body color and pattern to resemble their surroundings.

Unusual senses

As with snakes, chemicals in the environment are very important for lizards. They sense them by protruding the tongue, and then withdrawing it and inserting it into a hollow gland at the top of the mouth, called Jacobson's organ. Many lizard species use this sense to give them information about potential food. It is now becoming clear that they use it for many other things, too, for example to determine the sex and sometimes the identity of another individual, or to determine whether a potential predator has passed by recently. There have been other recent discoveries about the senses of lizards. Some species can detect ultraviolet wavelengths, as many birds can. This is important for understanding lizard behavior, because it means that what a lizard sees is not necessarily the same as what a human observer sees when looking at the same object.

Enigmatic eye

Most lizards have an extra eye on the top of the head, but it is located beneath the skin so it is not visible without dissection. However, because the scale above this eye is slightly thinner than elsewhere and has less pigment, the eye's position can usually be discerned. Like the

Lizards Family Tree

Order	Squamata				
Suborder		Sauria			
Division		Ascalabota		Autarchoglossa	
Superfamily	Dibamia	Gekkota	Iguania	Scincomorpha	Anguimorpha
Family	Dibamidae	Gekkonidae Pygopodidae Xantusiidae	Agamidae Chamaeleontidae Iguanidae	Cordylidae Lacertidae Scincidae Teiidae	Anniellidae Anguidae Helodermatidae Lanthanotidae Varanidae Xenosauridae
Species	*Blind lizards*	*Geckos Snake lizards Night lizards*	*Chameleons Agama lizards (including bearded lizards, frilled lizards, moloch and flying lizards) Iguanas (including anole lizards, basilisks, chuckwallas, horned lizards and spiny lizards)*	*Skinks Plated lizards Zonures Wall lizards (including green lizard, sand lizard and viviparous lizard) Teiid lizards (including whiptails and tegus)*	*California legless lizards Glass lizards Slowworm Alligator lizards Monitor lizards (including Komodo dragon) Gila monster Beaded lizard Earless monitor Chinese crocodile lizard*

Many of the approximately 1,000 species of skinks are legless, such as this burrowing skink from Australia. Legless lizards are well adapted to life in desert sand, underground or in thick vegetation.

Knob-tailed geckos, Nephruros asper, *live in rocky habitats in Australia. The function of their bizarre tails is unknown.*

familiar lateral eyes of vertebrates, it has a retina and a lens. It is called the parietal eye. This eye cannot produce an image, but experiments have shown that it can discriminate the intensity of light and also its color.

There has been a great deal of debate as to what the parietal eye is for, although it probably has a number of functions. It helps the lizard to know the time of day and to sense the passing of the seasons. However, it is not essential for this, as some diurnal lizards and most nocturnal lizards do not have the parietal eye. It is obviously part of the system that controls seasonal breeding, but the eye's precise role is not known. It is thought that it may help the lizard to control its body temperature. There has even been a suggestion that, at least in some species, the parietal eye may determine how light is polarized, and so help a lizard to navigate and not get lost.

Living young

Most lizard species lay eggs. Although some species have hard-shelled eggs, like birds, in most the shell has a parchmentlike texture. In a few species, however, the eggs have only a very thin, membranous shell, and they are retained in the uterus of the female until they are ready to hatch. In this case the mother gives birth either to living young that have already hatched within the uterus, or produces young that hatch immediately from within a membranous covering. Lizards that reproduce in this way are called ovoviviparous. While ovoviviparous species can be found in many parts of the world, they are particularly associated with cool climates.

This is because, by moving from place to place, the mother can make sure that the developing embryos are always in the warmest part of the habitat, and so will develop most rapidly. In a small number of ovoviviparous species there is some transfer of material across the wall of the uterus to the developing embryos. This is a primitive kind of placenta.

Sun lovers

Lizards, in common with all reptiles, cannot control their body temperature by retaining metabolic heat (the heat produced by burning food) as mammals and birds are able to do. Individuals captured during the daytime, however, will often feel hot to the touch. The reason for this is that they have used the warmth of the sun to raise their body temperature above that of most of their surroundings. There are many kinds of animals that do this, including some frogs, butterflies, beetles, dragonflies and bees. Lizards, however, can often maintain high temperatures with high levels of accuracy for long periods of time.

When lizards need to warm up, they place themselves in direct sunshine, or on sun-warmed substrates such as hot rocks, bark or soil. When they get too hot, they simply move away into shade. In extremely hot environments, such as deserts in the Tropics, they may pant or crawl up into bushes or onto rocks to catch any cooling breezes. Collectively, behavior patterns that are designed to help animals control their

A wavy crest of elongated scales runs right along the back of the common iguana, Iguana iguana. Many iguanas, agama lizards and chameleons possess flaps, frills, crests or other ornamentation, used in defense and displays.

time available to do other things, such as looking for food or finding a mate. Eventually, however, as the sun sinks during the late afternoon and temperatures begin to fall, basking periods will increase again. Thermoregulating lizards are utterly dependent on the sun.

Lizards in densely vegetated habitats, such as rain forest, may have to spend a lot of time looking for patches of sunshine to bask in. Therefore some species are not able to thermoregulate. Nocturnal lizards cannot usually thermoregulate, although geckos may seek out sun-warmed surfaces when they first become active each evening.

Each species of lizard has a characteristic temperature that it attempts to maintain. The selected temperature is known as the mean activity temperature. In many species this is more-or-less equivalent to human blood heat. In desert lizards this temperature is often higher, although it is only very rarely above 102–104° F (39–40° C). In burrowing lizards, or those that live in thick vegetation, the mean activity temperature is less than human blood heat.

body temperature are called behavioral thermoregulation. Such behavior patterns are particularly important for many lizards, and appreciating how and why they work can help to provide an understanding of many aspects of their lives.

In temperate climates, behavioral thermoregulation determines when individuals are active and when they are at rest. Lizards sometimes remain basking in the sun for long periods while they are warming up. Although they remain alert, they are easiest to find and capture at this time because they are motionless and exposed to view. As the day progresses, periods spent basking will decrease. An individual will have more

For particular species see:
- AGAMA LIZARD • ANOLE LIZARD • BASILISK
- BEARDED LIZARD • CHAMELEON • CHUCKWALLA
- EARLESS MONITOR • FENCE LIZARD • FRILLED LIZARD
- GECKO • GILA MONSTER • GLASS LIZARD
- GREEN LIZARD • HORNED LIZARD • IGUANA
- KOMODO DRAGON • LEGLESS LIZARD
- MARINE IGUANA • MOLOCH • MONITOR LIZARD
- NIGHT LIZARD • SAND LIZARD • SKINK
- SLOWWORM • TEIID LIZARD • TUATARA
- VIVIPAROUS LIZARD • WALL LIZARD

LLAMA

It is thought that the domesticated llama was bred from its wild relative the guañaco during or before the time of the Inca civilization. Then, as now, it was mainly used as a pack animal.

THE LLAMA, A SOUTH AMERICAN member of the camel family, Camelidae, has long been domesticated, as has its close relative, the alpaca, *Lama pacos*. The wild ancestor of both the alpaca and the llama, *L. glama*, is thought to be the guañaco, *L. guanicoe*. The other non-domesticated member of the group is the vicuña, *Vicugna vicugna*. The four species are collectively known as lamoids and while they resemble camels in many respects, all are smaller and do not have the camels' characteristic hump. They have long legs, erect ears and thick muzzles. The domestic species are distinctive for their very thick, woolly coats.

The llama's wild relatives

The guañaco is found from southern Peru to the southern plains of Patagonia. It is about 6 feet (1.8 m) in head and body length, including a fairly long neck, and its tail can be up to 10 inches (25 cm) long. It stands between 3 and 4 feet (0.9–1.2 m) at the shoulder, and weighs about 260 pounds (120 kg). The guañaco's woolly coat is tawny to brown in color and its head is gray. If threatened or disturbed, it is able to run at speeds of up to 40 miles per hour (65 km/h).

The vicuña is only slightly smaller than the guañaco in head and body length, but has a slighter build and weighs considerably less, averaging just 100 pounds (45 kg). It ranges from northern Peru to the northern parts of Chile, keeping more to the mountains, usually at altitudes of 11,500–20,000 feet (3,500–6,000 m). It stands about 3 feet (0.9 m) at the shoulder and has a light brown coat and yellowish red bib.

Early domestication

When in 1531 the Spanish conquistadors overran the Inca Empire in the high Andes, they found the llama and alpaca already numerous as domesticated animals. The llama is the largest of the lamoids, reaching more than 7 feet (2.2 m) in

LLAMA AND RELATIVES

CLASS	**Mammalia**
ORDER	**Artiodactyla**
FAMILY	**Camelidae**

GENUS AND SPECIES **Llama, *Lama glama*; guañaco, *L. guanicoe*; alpaca, *L. pacos*; vicuña, *Vicugna vicugna***

WEIGHT
Llama and guañaco: 220–330 lb. (100–150 kg). Vicuña and alpaca: 75–145 lb. (35–65 kg).

LENGTH
Llama and guañaco, head and body: 4–7¼ ft. (1.2–2.2 m); shoulder height: up to 4 ft. (1.2 m); tail: 6–10 in. (15–25 cm). Vicuña and alpaca, head and body: 4–6¼ ft. (1.2–1.9 m); shoulder height: up to 3 ft. (0.9 m); tail: 6–10 in. (15–25 cm).

DISTINCTIVE FEATURES
Resemble small, humpless camels with erect ears, thick muzzles and long legs. Llama and alpaca (domestic): very thick, woolly coats. Vicuña (wild): slighter build; longer, thinner neck. Guañaco (wild): like a large vicuña.

DIET
Grasses and herbs; leaves where available

BREEDING
Breeding season: November–May (llama); number of young: 1; gestation period: 345–360 days; breeding interval: 1 year (vicuña), 2 years (llama)

LIFE SPAN
Up to 28 years in captivity

HABITAT
High mountain grasslands and alpine areas

DISTRIBUTION
Peru south to Argentina and Chile

STATUS
Llama and alpaca: numerous in captivity or semiwild conditions. Guañaco: uncommon and declining. Vicuña: endangered.

The vicuña, one of the two nondomesticated relatives of the llama, is now endangered. This vicuña herd is browsing on water plants in Lanca National Park, Chile.

length and weighing up to 330 pounds (150 kg). It has a long, dense, woolly coat. It was and still is used mainly as a beast of burden. Although it is a gentle animal, the llama will spit, kick and bite if it is overloaded or mistreated.

The alpaca, smaller than the llama, was selectively bred for its coat, which makes a finer quality wool than that of any other animal. Its coat is also much longer than that of the llama and was formerly woven into robes used by Incan royalty.

Social animals
Lamoids live in herds comprising one male and a few females with their young. They mainly graze on grass and but will browse the leaves from trees and shrubs when available.

Mating takes place in the llama from November to May. The single young is born 345–360 days later. The young llama is precocial (capable of a high degree of independent activity), being able to run swiftly, for example, soon after birth. It is weaned at around 6–12 weeks. The llama breeds only every 2 years, but the vicuña is known to breed annually.

Which ancestor?
Bones that match those of both the llama and the alpaca have been unearthed in human settlements, suggesting that their domestication goes back 4,500 years. The animals may have been domesticated even earlier than this, so we can only guess at their wild ancestors. In this there is a difference of opinion among zoologists. The first study was made by the Austrian paleontologist

G. Antonius who, in 1922, argued that the llama was derived from the guañaco and the alpaca from the vicuña. Thirty years later more zoologists reexamined the arguments and decided a study of the animals' skulls showed that the two domesticated forms were both derived from one wild ancestor, the guañaco. Both schools of thought continue to have their followers today.

The matter has not been settled by breeding. The llama and the alpaca interbreed to produce fertile offspring, strongly suggesting that they belong to one species. However, in captivity both the guañaco and the vicuña also interbreed to give fertile offspring. This is unusual if the two are distinct species. There are other views on the matter. One is that the alpaca may have been derived from hybrids between the wild vicuña and the domesticated llama. The other is that possibly both llama and vicuña were bred from a wild species that long ago became extinct.

Life at the top

Most llama herds are kept by the native peoples of Bolivia, Peru, Ecuador, Chile and Argentina. Although today the motorized vehicle tends to oust the llama as a pack animal, the species is still of prime importance to these people living high in the Andes. The llama and other lamoids are well suited for life at these heights because their hemoglobin (the respiratory pigment of red blood cells) can take in more oxygen than that of other mammals. Their red corpuscles also have a longer life span: 235 days as compared to the 100 days of human blood corpuscles. The llama is also suited to its domestic role in that it is able to subsist on a wide variety of vegetation and can go for long periods without water. Apart from its use as transport, the llama provides meat, wool for clothing and rugs, hides for sandals and fat for candles. Its wool when braided is also used for ropes, and its dung is dried and used for fuel.

Persecution for tourism

Wild lamoids have long been corralled, sheared and then released, and the development of tourism in the Andes Mountains has created a growing demand for products from these animals. One result is a heavy persecution of the wild species so that they are becoming rare. The guañaco is declining, and the vicuña is now endangered. It is thought that only about 150,000 vicuña remain in the wild.

The guañaco, able to absorb more oxygen into its blood, is well suited to its existence high in the Andes.

LOACH

LOACHES AND SPINY LOACHES live in the fresh waters of Europe and Asia, including the Malay Archipelago, as well as in isolated areas of Morocco and Ethiopia. There are 110 species, grouped into 18 genera. Loaches are small fish, seldom more than 15¾ inches (40 cm) long, and while a few have a familiar fish shape, most are slender-bodied apart from a flattened undersurface. Most loaches have a conspicuous pattern of dark bands on the body, though it may be broken up until often only a marbling remains. Loaches have small scales in the skin. There are three to six pairs of barbels around the toothless mouth, which is set back beneath the snout. The single dorsal fin, the anal fin and the paired fins are all much the same size.

Eurasian species
Most well-known species of loach, such as the tiger loach, *Botia hymenophysa*, and the coolie loach, *Pangio kuhlii*, are native to southern and Southeast Asia. Exceptions include the stone loach (*Barbatula barbatula*), the weatherfish (*Misgurnus fossilis*) and the spined loach (*Cobitis taenia*). The latter is found in Europe and northern Asia and is named for a spine, which may fork at its tip, protruding below and in front of the eye. The spine occurs on several loach species. It may be erected and fixed and so become fatal to predators. Larger fish or birds may perish by swallowing loaches and getting the spines stuck in their gullets. The stone loach ranges from Ireland, where it was introduced, across Europe, Asia and China.

The weatherfish is distributed in central and eastern Europe. It derives its name from its reputed ability to anticipate changes in the weather. Several loaches are said to grow restless before a thunderstorm, and the name loach allegedly derives from the French *locher*, "to fidget." One theory dating back to 1895 suggests that bones connecting the loach's swimbladder with its inner ear may enable the fish to detect changes in barometric pressure. Both the weatherfish and the spined loach are protected by law.

Burrowing for food
Loaches are bottom-living fish. Those living in streams have rounded bodies, somewhat flattened from the top down in some species, while those living in still water have bodies flattened

from side to side. Loaches use their barbels to hunt for food, mainly insect larvae and worms. Some species burrow into the sand or mud at the bottom, either to escape predators or to pass the winter. Loaches also eat algae and at one time were often kept by aquarists to clear aquarium walls of algal coatings. Their usual method of feeding, however, is to comb the surface of the sand or mud, swallowing edible particles and passing solid grains out through the gills.

Two ways of breathing
Loaches have gills and a swim bladder that, in some species at least, is enclosed in a bony capsule. Many loaches breathe through the intestine. If the water becomes impure, they come to the surface and gulp air, which they swallow. They also gulp air if their pond dries up. In this event, the loaches bury themselves in the mud and wait for rain. The wall of the hind end of the intestine is rich in small blood vessels and works rather like a lung, taking in oxygen and giving out carbon dioxide, the spent air being expelled through a vent.

Experiments in aquariums show a direct link between the loaches' habit of gulping air and rising water temperature. The warmer the water, the less oxygen it holds. At 41° F (5° C) a loach breathes solely through its gills. At 50° F (10° C)

Loaches are popular aquarium fish. Pictured is the most colorful species, the clown loach, Botia macracanthus.

Also called the green tiger botia, the tiger loach remains hidden by day. It emerges at nightfall to burrow for insect larvae and worms.

a bubble of air is swallowed every 2 hours. At 59° F (15° C) five bubbles are swallowed in an hour, rising to 10 bubbles per hour at 77° F (25° C). If, however, a loach is placed in water at 77° F (25° C) from which the oxygen has been removed, it will swallow 70 bubbles an hour, coming to the surface every minute or so to gulp.

Switching from gills to lungs

In some species of loach most of the intestine is used as a lung. This is true of the stone loach, spiny loach and weatherfish, as well as of an Indian species, *Lepidocephalus guntea*, and the Chinese weatherfish, *Misgurnus anguillicaudatus*. After gulping air, *L. guntea* turns a somersault, at the same time driving out the spent air through a vent. A stream of 8 to 12 bubbles is ejected with force, producing a distinct clicking sound. Scientific analysis of these bubbles reveals that they contain only a small percentage of carbon dioxide, the remaining gases being nitrogen and some oxygen. Most of the carbon dioxide is expelled through the gills.

The Chinese weatherfish is notable in that it has a seasonal lung. It uses gills only in winter, when there is more oxygen in the cold water, in summer becoming an air-gulper. Moreover, as the fish resumes its gill-breathing the "lung" part of the intestine reverts to digestive tissue.

Breeding

Although loaches have long been popular aquarium fish, little is known of the breeding habits of most species. They have only occasionally spawned in captivity, and accounts of what happens are conflicting. For example, reports vary as to whether the eggs are laid on sand or in a bubble nest. More is known about the stone loach, which has the reputation of being a prolific breeder. It spawns indiscriminately on

WEATHERFISH

CLASS	**Osteichthyes**
ORDER	**Cypriniformes**
FAMILY	**Cobitidae**
GENUS AND SPECIES	*Misgurnus fossilis*

LENGTH
Up to 12 in. (30 cm)

DISTINCTIVE FEATURES
Long, slender, eel-like body with small head; small dorsal fin, located closer to tail than head; 5 pairs of barbels around mouth; dark gray above, lighter on ventral (lower) surface; several dark stripes run lengthwise along body

DIET
Insect larvae and small mollusks

BREEDING
Age at first breeding: not known; breeding season: April–June; number of eggs: up to 150,000; hatching period: 8–10 days; breeding interval: 1 year

LIFE SPAN
Not known

HABITAT
Sandy or muddy bottoms of lowland ponds, pools, marshes, river backwaters and ditches

DISTRIBUTION
Central and eastern Europe

STATUS
At low risk

stones and pebbles in April–May, the eggs being laid at night. The eggs are large for such a small fish, 1 millimeter in diameter, and are sticky and numerous. They hatch in 8–11 days. After hatching the young lie on the bottom at first, beginning to feed 8–10 days later.

LOBSTER

OBSTERS ARE ANY of a number of marine crustaceans from several families. The true lobsters are from the superfamily Nephropsidea, but there are also spiny lobsters, family Palinuridae, and others. Most lobsters are nocturnal and all are bottom-dwelling, scavenging a wide variety of both living and dead animals.

The common or European lobster, *Homarus gammarus*, is very like some crayfish, but differs in being larger, up to 20 inches (50 cm) long and weighing up to 22 pounds (10 kg). It has a rigid, segmented body and is predominantly a mottled bluish color, but turns red when boiled. One pincer or claw is very large and heavy and is used for crushing, while the other is smaller and toothed for holding and tearing food. There are also small pincers on the first two of the four pairs of walking legs.

Numerous species

The American East Coast lobster, *H. americanus*, is like the European lobster in all but a few minor details. It is blackish green above and orange, red or blue below. The largest on record weighed 44½ pounds (20 kg) and its body alone was 2 feet (60 cm) long. However, on both sides of the Atlantic the average size of lobsters has diminished greatly as a result of being fished. Few now live long enough to approach this size.

The many other kinds of lobsters include, in European waters, the smaller Norway lobster, *Nephrops norvegicus*, or Dublin prawn. It is up to 8 inches (20 cm) long and a delicate flesh color, and is the source of scampi.

There are also the spiny lobsters, crawfish or langoustines, genus *Palinurus*, which grow up to 18 inches (45 cm). Spiny lobsters lack the large, pinching claws of the true lobsters but have protective spines on the body and long spiny antennae, which can be used to repel predators. At the bases of the antennae of the Californian spiny lobster, *P. interruptus*, of the American West Coast, are sound-producing organs that the lobster uses when molested. It is thought the sound produced might scare off fish.

Both the European lobster and the Norway lobster range along European coasts from Norway south to the Mediterranean. The European lobster lives on rocky coasts, venturing up the shore only occasionally between tidemarks, especially in summer. The Norway lobster is trawled from muddy bottoms in deeper water, while

West Indian spiny lobsters, **Palinurus argus.** *Found from Bermuda south to Brazil, this species is commercially important.*

Spiny lobsters lack the large pincers of the true lobsters. Pictured is Palinurus elephas, *of Europe's rocky coastlines.*

Palinurus elephas is a species of spiny lobster that is found further south, on rocky coasts. It is more important as food in the Mediterranean region than elsewhere. The American lobster is found in fairly deep water, to depths of 1,200 feet (365 m), from Labrador south to North Carolina.

Burrowers and scavengers

Lobsters move over the seabed on four pairs of walking legs, often propelled forward by the beating of the swimmerets (small swimming appendages) under their abdomens. In addition, to escape a predator they may dart suddenly backward by flicking their jointed abdomens and flipperlike tail fans forward under their bodies. The tail is also used for swimming.

The European lobster, which is active particularly at night, spends much of its time with its long, sensory antennae protruding from the safety of a rocky crevice or a burrow it has dug itself. It scavenges a variety of animal matter, both living and dead, such as the stale fish used to bait lobster pots. It also eats a small amount of seaweed and eelgrass. The food is shredded into tiny pieces by a pair of mandibles, helped by the three pairs of maxillipeds, which together make up its jaws. Food is further broken up in the gizzard.

Thousands of eggs

The female European lobster begins breeding when she is about 8 inches (20 cm) long, at about 5 years old. She then spawns once every two years, in July and August. In mating, she lies on her back while the male, using specially modified swimmerets, places a packet of sperms in a receptacle between the last pairs of her walking legs. The thousands of dark green eggs she lays are cemented onto her swimmerets and remain there, protected and aerated, for 10–11 months. She is then said to be "in berry." The law forbids the sale of females in berry.

EUROPEAN LOBSTER

PHYLUM **Arthropoda**

CLASS **Crustacea**

ORDER **Decapoda**

SUPERFAMILY **Nephropsidea**

GENUS AND SPECIES ***Homarus gammarus***

ALTERNATIVE NAME
Common lobster

WEIGHT
Up to 22 lb. (10 kg)

LENGTH
Up to 20 in. (50 cm)

DISTINCTIVE FEATURES
Four pairs of walking limbs; 2 large pincers or claws, one very much larger than other; jointed abdomen, curving over to flipperlike tail fan; bluish mottled color

DIET
Wide variety of living and dead animals; occasionally seaweed and seagrasses

BREEDING
Age at first breeding: about 5 years, when 8 in. (20 cm) long; breeding season: July–August; number of young: thousands; hatching period: 10–11 months; breeding interval: 2 years

LIFE SPAN
Usually up to 40–50 years, may be much longer in some cases

HABITAT
Rocky areas in shallow subtidal waters down to depths of 200 ft. (60 m)

DISTRIBUTION
European coasts from Norway south to the Mediterranean

STATUS
Common

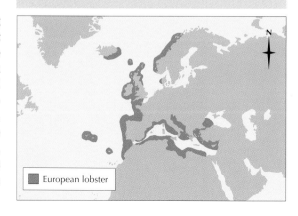

European lobster

Some idea of the numbers of eggs laid can be gained from the following figures for the American lobster. A female 8 inches (20 cm) long lays 5,000 eggs, one 10 inches (25 cm) long lays 10,000, and a female 1⅖ feet (43 cm) long lays 63,000. The largest number recorded was 97,440.

The young hatch as shrimplike "mysis" larvae about ⅓ inch (0.8 cm) long, with large eyes and no swimmerets. For the next few weeks or months they swim in the surface waters, then sink to the bottom when about ½ inch (1.3 cm) long.

Grow by molting

Young lobsters grow by periodic molting of the hard external skeleton. Molting takes place about eight times during the first year, less frequently during the next two years, and only once (female) or twice (male) each year after the third year. By the time lobsters are old there can be 2 years between molts, although (unlike some crustaceans) lobsters never cease molting altogether. Because of the long intermolt period, old lobsters become encrusted with barnacles and other marine life.

The skeleton is heavily impregnated with lime salts, and before a molt much of the lime is reabsorbed into the body for reuse. It is stored temporarily in the liver and as two large "stones" in the stomach. After a molt, a lobster develops a craving for calcium carbonate, which it may devour in the form of its own cast skin or the shells of other animals such as mollusks and sea urchins. If other sources are denied it, a lobster may turn cannibal. Even with adequate supplies of calcium carbonate, the shell takes about 6 weeks to become fully hardened, and during this time the lobster must remain in the protection of its home if it is not to fall prey to sharks, skates, dogfish and cod. However, the main threat to adult lobsters is commercial fishing.

Limbs regenerate

If a lobster is caught by one of its walking legs or more especially by a claw, it is likely that this will break off while the remainder of the animal escapes. It is not that the limbs are brittle, but there is a special reflex mechanism for sacrificing a trapped limb. The break occurs at a predetermined breaking plane visible as an encircling groove toward the base of the limb, and there is a special muscle that causes the fracture.

Limbs lost either in this way or by other means are replaced when the lobster next molts. Such limbs are replaced in miniature at first, but grow at successive molts until they reach normal size. Molting may occur sooner than usual, for example, within 150 days, but without the normal increase in body size, if more than three legs have been lost.

A European or common lobster. Mottled bluish, the familiar red color occurs only when the animal is immersed in boiling water.

LOCUST

solitary phase and a gregarious (swarming) phase. When the grasshoppers become crowded, there is a change in behavior. If such crowding continues for a generation or more, they also change their shape and color to that of the gregarious form.

Solitary locusts come together only for mating and then behave very much as other grasshoppers do. A solitary female desert locust lays 95 to 160 eggs about 4 inches (10 cm) below ground. The eggs hatch in 2–3 weeks. The larvae grow through a series of molts in the course of which they become gradually more like adult locusts. In the desert locust there are 5 or 6 larval stages, taking place over about 40 days. The flightless immature locusts are known as hoppers. In the final two molts the wings form, becoming functional when the insects become adults.

When they first hatch, young solitary locusts disperse and, unless some environmental factor forces them together, settle down to the solitary life of grasshoppers. These are usually green or brown, blending well with their surroundings.

Adult migratory locusts feeding on maize. Locust swarms or plagues can do enormous damage to commercial crops.

ALTHOUGH THE TERM locust is loosely applied to any large tropical or subtropical grasshopper, it is better restricted to those species the numbers of which occasionally build up to form enormous migrating swarms that may do catastrophic damage to vegetation, notably cultivated crops and plantations.

Africa suffers most seriously from locust swarms, and three species are of special importance. These are the desert locust (*Schistocerca gregaria*), the red-legged or red locust (*Nomadacris septemfasciata*) and the African subspecies of the migratory locust (*Locusta migratoria migratorioides*). For both the red locust and the African migratory locust there exist regional control organizations that effectively prevent plague outbreaks. The desert locust, however, continues to present a real international problem and will therefore be the main subject of this article.

Harmless, solitary phase

The vast plagues of locusts that periodically create such havoc in many regions of the developing world are really hordes of gregarious grasshoppers. Locusts can exist in two phases: a

Change to the swarming phase

Under certain conditions, depending on factors such as the weather, scattered solitary locusts may become concentrated into favorable laying sites. Females can lay at least three times in their lifetime, usually at intervals of about 6–11 days. With every female locust laying two or three batches, locust numbers can multiply rapidly. With many more hoppers crowded into the same area, groups coalesce, and they are then well on the way to the phase change from solitary to gregarious, which leads eventually to locust swarms.

When crowded together, the hoppers change in color to bold patterns of black and orange or yellow stripes. When they grow into adults they are pink at first, but after a few weeks or months, usually coinciding with the rainy season, they mature sexually and become yellow. This compares to the usual brown color of the solitary adult. In the adults there are also structural differences between the solitary and gregarious phases, notably in the wings, which are relatively longer in the gregarious phase. The structural changes between one phase and another are associated with differences in the insects' hormonal balance.

LOCUSTS

PHYLUM **Arthropoda**

CLASS **Insecta**

ORDER **Orthoptera**

FAMILY **Acrididae**

GENUS AND SPECIES **Desert locust, *Schistocerca gregaria*; migratory locust, *Locusta migratoria*; red-legged locust, *Nomadacris septemfasciata*; Australian plague locust, *Chortoicetes terminifera*; others**

ALTERNATIVE NAME
Red-legged locust: red locust

LENGTH
Migratory locust: about 2⅓ in. (6 cm)

DISTINCTIVE FEATURES
Large grasshoppers with strong wings. Desert locust: green or brown (solitary phase); pink (immature gregarious phase); yellow (adult gregarious phase); wings much longer in gregarious phase.

DIET
Plant materials

BREEDING
Desert locust, number of eggs: 95 to 160 (solitary phase) or up to 80 (gregarious phase); hatching period: 14–21 days; larval period: about 40 days

LIFE SPAN
Desert locust: 3–5 months

HABITAT
Desert locust: semiarid country and deserts (solitary phase), all available land habitats (gregarious phase)

DISTRIBUTION
Northern half of Africa east to the Middle East and southern Asia

STATUS
Common most of the time; periodically superabundant, forming vast swarms

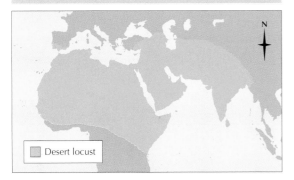

Desert locust

Mutual stimulation leads to greater activity and the locusts start to march. Because an urge to keep close together is induced by development in the gregarious phase, they march together in bands. As adults, the locusts continue to migrate, and the urge to crowd together is maintained, but they are now airborne and move much faster. As the region inhabited by the desert locust is mainly arid, the result of persistently flying with the wind is to concentrate large numbers of locusts in areas when rain is likely to fall and provide them with food.

Locusts are able to stay in the air for long periods of time and swarms can travel 3–80 miles (5–130 km) or more in a day. In the past there have been some spectacular and very long distance swarm migrations. For example, in 1954 a swarm traveled from northwestern Africa to the British Isles. Another swarm migrated from West Africa to the Caribbean, a distance of 3,120 miles (5,000 km), in about 10 days in 1988. Solitary desert locust adults usually fly at night, whereas the gregarious adults fly during the day.

Locust plagues

When the swarm descends, the locusts devour everything green. After mating takes place each female lays around 80 eggs. The eggs are held together by a frothy secretion that hardens to form an egg pod. Up to 1,000 egg pods have been found within 11 square feet (1 sq m).

The next generation of young are still crowded and the swarm that results can be many times larger than the original one. This swarm continues to migrate and again descends and multiplies its numbers. In this way a swarm may cover vast areas. Locust swarms can vary from less than 1 square mile to several hundred square miles. There can be at least 100 million and sometimes as many as 200 million locust adults in

When crowded together, desert locust hoppers (subadults) develop bright patterns of black and orange or yellow stripes.

each square mile. Plagues eventually come to an end because of adverse weather conditions. The few survivors revert to the solitary phase.

Perhaps the most important discovery made by the Anti-Locust Research Centre in London has been that the African migratory locust and the red-legged locust change from the solitary to the gregarious phase only in certain limited outbreak areas. These areas are effectively policed by regional locust control organizations, and plague outbreaks have been prevented since 1944.

The desert locust is the locust of the Bible and is now by far the most damaging of all. It is hard to control because it has no geographically determined outbreak areas. It ranges over a vast area, from southern Spain and Asia Minor, south to the whole of northern Africa and east through Iran to Bangladesh and India, an area comprising about 60 countries. Between plagues lasting 6 or more years there are equally long recessions when only solitary locusts are to be found. The last plague ran from 1986 to 1989.

Natural predators

Fortunately, there is a high mortality of locusts from natural causes. Winds may fail to carry the locusts to a suitable breeding area. The soil may not be moist enough for the eggs to hatch. Or they hatch to find insufficient plant growth for their food or to protect them from the heat of the midday sun. There are also many predators, parasites and diseases of locusts. However, thus far control by natural predators and parasites is limited since locusts can quickly migrate away

from unfavorable conditions and most predators. People and birds often eat locusts but usually not enough to significantly reduce population levels over large areas.

Control by humans

Insecticides, mainly organophosphate chemicals applied in small doses, are used in campaigns against desert locusts. These are sprayed both from the air and from ground vehicles. Regional locust control organizations within the desert locust invasion areas pool resources and facilitate the movement of supplies across frontiers. These organizations in turn are coordinated by the Food and Agriculture Organization (FAO) of the United Nations, with its headquarters in Rome. The FAO Emergency Prevention System is a program that aims to minimize the risk of emergencies, such as the desert locust plague of 1986–1989, developing. The program is initially concentrated in nine countries that border the Red Sea and Gulf of Aden, which have historically been the source of many outbreaks and plagues. The program is also being extended to West Africa.

However, there are many reasons as to why it is difficult to successfully combat the desert locust. These include difficulty in predicting outbreaks, the extremely large area within which locusts can be found, the remoteness and difficulty in accessing such areas and the limited resources for locust monitoring and control in some of the affected countries. Other problems include wars, lack of infrastructure and poor political relations among affected countries.

Young migratory locusts in South Africa. If conditions lead to solitary-phase locusts being concentrated in favorable laying sites, massive swarms of gregarious-phase locusts may form.

LOGGERHEAD TURTLE

THE LOGGERHEAD IS ONE of the sea turtles, distinguished from the green turtle, *Chelonia mydas*, by its larger head and by the arrangement of the plates on the carapace, or shell. The loggerhead turtle is olive or reddish brown above and the shields are often edged with yellow, especially at their lower edges. It is yellow or cream colored on the underside of the shell. The loggerhead is the second largest of the marine turtle species, exceeded in size only by the leatherback turtle, *Dermochelys coriacea*. The biggest loggerhead ever recorded probably weighed about 1,160 pounds (540 kg). However, most mature adult specimens are within the range of 215–320 pounds (100–150 kg), reaching about 3 feet (0.9 m) in length. Young loggerheads have three keels running along the carapace, but the outer ones disappear as the turtle ages and even the remaining middle keel gets progressively less pronounced.

Widespread distribution

The loggerhead is the only species of marine turtle to breed on the coasts of mainland United States, in South Carolina, Georgia and Florida. Elsewhere nesting takes place in the Caribbean and Central and South America, in the Mediterranean, around the coasts of southern Africa, off southern Arabia and in Japan, Fiji, the Solomon Islands and western Australia. When they are not breeding, loggerhead turtles wander the Pacific, Indian and Atlantic Oceans. They do not go so far into the mid-ocean as the leatherback turtle, and are rarely found more than 150 miles (240 km) out to sea. They sometimes wander into estuaries, and have been seen in the Mississippi River. It has also been discovered that some individuals hibernate on the seabed off the east coast of Florida.

Mainly predatory

The loggerhead turtle feeds mainly on fish and marine invertebrates, including floating crustaceans, mollusks and jellyfish. However, it will also eat seaweed and turtle grass. Like the hawksbill turtle, *Eretmochelys imbricata*, the loggerhead eats the very poisonous Portuguese man-of-war, attacking with its eyes closed. It is assumed that turtles are immune to the poison of this and of the other jellyfish they eat with relish.

Repeated egg-laying visits

The timing of the breeding season varies from place to place. On the eastern seaboard of the United States, for example, loggerhead turtles nest from April to August. An individual female may come ashore up to three times during a season to lay eggs. After that she will rest for 2–3 years before nesting again.

Loggerhead turtles are easily disturbed when they come ashore to lay their eggs. Disturbance from tourism on the nesting beaches is one of the factors that has lead to the vulnerability of this species.

Unlike leatherback turtles, which seem to prefer open beaches, loggerhead turtles nest on beaches guarded by rocks and reefs, where they feed before coming ashore. Loggerheads crawl about 300 feet (90 m) up the shore, farther than the leatherbacks. Their nesting habits are otherwise very similar. The female digs a body pit, throwing sand sideways until she is lying in a hole about 6 inches (15 cm) deep. Then, after a short rest, the egg pit is excavated, lifting out cupfuls of sand with the hind flippers until it is 8 inches (20 cm) deep. While crawling up the beach, turtles are very easily disturbed and will retreat into the waves, but once they have started to dig they continue until the job is done.

Egg-laying takes 10–30 minutes and the eggs are dropped three or four at a time. The clutch averages around 120, the maximum being 200. On completing the clutch the pit is filled in with sand and the sand is scattered widely so that a large area appears churned and the actual nest is difficult to find. The eggs hatch after 8–10 weeks, the baby turtles digging themselves out of the sand together and rushing down the beach.

Threat of tourism

Despite their widespread distribution, loggerhead turtles are very vulnerable. All except their most remote nesting beaches are subject to increasing disturbance from tourism. This problem is particularly acute in the United States and in the Mediterranean. Even when the beaches are guarded from disturbance during nesting times, lights from the shore disorient the animals and they may not breed successfully. In many parts of the world their eggs are dug up and eaten in large numbers, by people, pigs, raccoons (an increasing problem in the United States) and other predators. Many adult turtles are also drowned in nets used for capturing fish and shrimps.

A loggerhead turtle hatchling, showing the three distinctive keels running along the carapace, or shell. Many young are taken by predators such as seabirds before they reach the relative safety of the open sea.

LOGGERHEAD TURTLE

CLASS	**Reptilia**
ORDER	**Testudines**
FAMILY	**Cheloniidae**
GENUS AND SPECIES	***Caretta caretta***

WEIGHT
Usually 215–320 lb. (100–150 kg)

LENGTH
Up to 3 ft. (0.9 m)

DISTINCTIVE FEATURES
Heart-shaped carapace (shell); broad head; strong, horny beak. Adult: olive brown or reddish brown above, often with yellow edge to carapace; yellow or cream underside; single keel (ridge) runs along middle of back, becoming less distinct with age. Young: 3 keels run lengthwise along back.

DIET
Fish and marine invertebrates such as jellyfish; also seaweed and turtle grass

BREEDING
Age at first breeding: probably 12–20 years; breeding season: April–September (Atlantic and Mediterranean), October–February (South Africa), October–April (Australia); number of eggs: average clutch 120; hatching period: 56–80 days; breeding interval: female lays up to 3 clutches in a season, once every 2–3 years

LIFE SPAN
Not known

HABITAT
Open seas, mainly in warm regions; nests on sandy beaches

DISTRIBUTION
Pacific, Indian and Atlantic Oceans (including the Mediterranean)

STATUS
Vulnerable; many populations declining

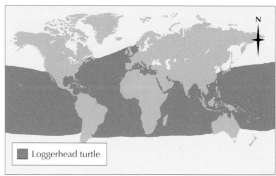

☐ Loggerhead turtle

LONG-HORNED BEETLE

THE NAME LONG-HORNED beetle is used to describe the beetles of the family Cerambycidae, which number about 15,000 species. Also known as longicorns or wood-boring beetles, the great majority are tropical species, although some are found in temperate regions.

The most obvious feature of long-horned beetles is their very long antennae, to which the name longhorn refers. An extreme form is seen in the members of the genus *Acanthocinus*, in which the antennae are four or more times the length of the body. One of these species, *A. aedilis*, is found across Europe and in North America.

Some species very large

Long-horned beetles vary greatly in size, from ½–6 inches (1.5–15 cm), excluding the antennae. Some of the tropical species are among the largest insects. For example, *Xixuthrus heros*, from Fiji, is 6 inches (15 cm) long, with the antennae being as long again. The New World species of *Macrodontia* are equally large in the body but have rather short antennae and enormous spiky jaws, rather like those of the unrelated stag beetle, *Lucanus cervus*, of the warmer regions of Europe.

Long-horned beetles are of great economic importance as pests of timber. The larvae of many species burrow in living or seasoned wood, making large tunnels that weaken it and spoil it for structural purposes.

Creaking beetles

Adult long-horned beetles are usually found on or near the trees in which the larvae feed, often hiding under loose bark. The adults themselves feed on pollen and are frequently seen feeding and sunning themselves on wild flowers. Many species fly well, and they may be attracted to lights at night.

Some of the large tropical long-horned beetles bite if handled, and there are also species that make a creaking noise. Known as stridulation, this is undoubtedly a defense reaction. These beetles make this sound either by moving the thorax (the middle of the three body sections found in insects) up and down, producing friction between it and the abdomen, or by scraping the hind legs against the edges of the elytra (wing cases).

Exceptional wood eaters

The larvae of most long-horned beetles feed inside the stems of plants and often inside trees, either under the bark or burrowing in the solid wood. Most species are found in one or a few kinds of tree. Some live exclusively in pine, for example, while others are found only in oak, apple or poplar. The eggs are laid on the bark and the tiny larvae burrow in. As they grow, they develop into large white or yellowish grubs with round heads and extremely powerful jaws with which they can rasp away the hardest heartwoods.

Most insects that feed on wood cannot directly digest the cellulose: they must either devour great quantities of wood, like the caterpillar of the goat moth, *Cossus cossus*, to get the noncellulose protein, or they have bacteria or protozoans in the alimentary canal that break down the cellulose, as in termites and the stag beetle. The grubs of long-horned beetles are exceptional in having a digestive enzyme that breaks down cellulose.

The larvae of most long-horned beetles take 2–3 years to reach full size and some take a good deal longer. When nearly ready to change into the pupa, the larva makes a tunnel to the exterior and then stops it up with a plug of wood fibers or a cap of hardened, chalky mucus. When the

Long-horned beetles are named for their exceptionally long antennae. These can be four or more times longer than the body in species of the genus Acanthocinus, such as A. aedilus (above).

beetle emerges from the pupa, it pushes or bites its way out. The adult beetle is unable to gnaw its way through solid wood, but it can bite through the plug made by the larva.

The grubs that live in the heartwood are fairly safe from predators, but those under bark or in decayed wood are a favorite prey of woodpeckers. There is a large long-horned beetle found in New Zealand, the grubs of which, known by the name of *hu-hu*, were traditionally regarded as a delicacy by the Maori people.

Camouflage and mimicry

The long-horned beetles are varied in color and marking, and some exhibit adaptive coloration. For example, the African species *Pterognatha gigas* is colored and mottled to look like a patch of moss or lichen, its long antennae sticking out like strands. Just as remarkable are the many cases of long-horned beetles that have come to resemble other insects that are distasteful or poisonous and so are left alone by predators. Many examples of mimicry of this kind are known from the Tropics. There is one common British species, the wasp beetle, *Clytus arietis*, the coloring of which mimics a wasp. It also behaves in a wasplike manner by scuttling in a agitated way over tree trunks and vegetation and by tapping its antennae like a wasp. Some tropical species of *Clytus* mimic ants in their appearance.

Some long-horned beetle species mimic distasteful or stinging insects, such as wasps, to deter predators from attacking. Pictured is Plagionotus detritus.

Town and country pests

The larvae of the house longhorn, *Hylotrupes bajulus*, live in dry, seasoned softwoods and are a serious pest of structural timber in Europe. The beetle is grayish black with two paler gray marks across the wing cases. The larva grows to 1 inch (2.5 cm) in length, and in warm weather the rasping of its jaws as it feeds is distinctly audible. A slow-growing grub, it seldom reaches full size in less than 3 years and may live for 10 years.

A pest found closer to home is the Asian long-horned beetle, *Anaplophora glabripennis*, a black-and-white spotted species that came to the United States in shipping containers from China. Its larvae tunnel through trees and, despite costly extermination programs, have caused millions of dollars' worth of damage, from Central Park in New York to rural lumber plantations.

LONG-HORNED BEETLES

PHYLUM	**Arthropoda**
CLASS	**Insecta**
ORDER	**Coleoptera**
FAMILY	**Cerambycidae**
GENUS	***Macrodontia, Acanthocinus, Clytus, Plagionotus, Hylotrupes, Strangalia, Aromia, Rhagium, Prionus, Desmocerus*; many others**
SPECIES	**About 15,000 species**

ALTERNATIVE NAMES
Adult: longicorn; wood-boring beetle; timberman. Larva: round-headed borer.

LENGTH
½–6 in. (1.5–15 cm)

DISTINCTIVE FEATURES
Adult: extremely long antennae; varied colors and markings; some species mimic other insects such as wasps and ants. Larva: rounded head; strong jaws; yellowish or whitish in color.

DIET
Adult: pollen. Larva: wood and plant stems.

BREEDING
Holometabolous (goes through a complete metamorphosis). Larval period: 2–3 years.

LIFE SPAN
Usually several years; up to 10 years in some species

HABITAT
Often wooded habitats, but wherever there is suitable feeding material for larvae

DISTRIBUTION
Mainly Tropics; smaller number of species in temperate regions

STATUS
Generally common

LOON

THE LOONS, OR DIVERS, are water birds with streamlined bodies, very short tails and straight, pointed bills. Their legs are set well back, like other powerful swimmers such as grebes, darters and cormorants. In common with grebes, but unlike darters and cormorants, loons have not developed an upright stance on land. Furthermore, their legs are enclosed in the body down to the ankle joint, so loons can only shuffle clumsily, a few paces at a time.

"Loon" is a North American name derived from the Old Norse *lómr*, or awkward person, in allusion to the birds' shuffling gait on land. In Iceland they are still known as lomr.

Northern breeders
When breeding, loons are distributed around the higher latitudes in the Northern Hemisphere. The common loon or great northern diver, *Gavia immer*, has a black and white spotted body and collar in summer. It breeds in most of Canada, extending south into the northern parts of the United States, as well as in Greenland, Iceland and Bear Island. In the winter, it flies to rocky coasts and generally moves southward. The yellow-billed loon, *G. adamsii*, known as the white-billed diver in Britain, closely resembles the common loon except for its pale, whitish yellow bill.

The Arctic loon or black-throated diver, *G. arctica*, and the red-throated loon or red-throated diver, *G. stellata*, have similar plumage patterns and distributions. In summer the former has black and white barrings on its back, dark gray on the head and neck and a black patch on the throat. In the red-throated loon the throat patch is reddish chestnut and the head and neck are dove gray. Both species breed in northern Canada, Siberia and northern Europe, including Scandinavia and Scotland. The red-throated loon is also found on Greenland, Iceland, the Faeroes, Spitzbergen and other Arctic islands.

Powerful swimmers
Loons are strong swimmers, and by expelling air from their bodies and plumage, they can swim with only head and neck showing. They may submerge gradually, so quietly that hardly a ripple is left, but at other times they plunge straight downward. During the breeding season,

loons are found on ponds, lakes and slow-moving rivers. Common loons favor remote, large or medium-sized lakes with plenty of undisturbed deep water. Red-throated loons, on the other hand, prefer much smaller, shallower pools and can therefore occur at greater densities than the other species.

Diving for fish
While feeding, loons stay under for a minute at the most, most dives lasting only 10–20 seconds. If alarmed, however, they can stay under longer and dive to great depths. One was caught in a fishing net below 200 feet (60 m). Loons swim with their partially webbed feet, using their wings only as stabilizers.

Loons eat mainly fish, both freshwater and marine, including species such as sand eels, gudgeon and gobies and the young of larger fish such as flounders, trout and perch. Some fish eggs, crustaceans, mollusks, aquatic insects and frogs are also eaten.

Sometimes fishers complain that loons spoil the fishing, but even when the birds gather in loose flocks in winter, they are very unlikely to have any serious effect on fish populations. It is true that loons do on occasion become tangled in lines and nets, damaging them.

Loons (Arctic loon, above) are superbly adapted for swimming fast and chasing fish underwater. They make strange wailing or whistling calls in the breeding season.

A common loon on its simple waterside nest. Loons find moving on land so difficult that they nest right at the water's edge.

Loons can take off only from the water, and then only after a long run across the surface. The wings are small for the size of the body, but loons fly powerfully once aloft and have been clocked at 60 mph (97 km/h). Landing is spectacular; the loon circles the pool or lake on rapidly beating wings, then glides down at a steep angle. As it skims toward the water it lowers its feet and slides to a halt with spray thrown up on either side.

The cry of the northlands

No one who has heard the eerie call of a loon is likely to forget it; it is a symbol of the bird's wild northern homeland. It is impossible to describe this call except poetically. The red-throated loon has been described as calling like the "wail of a lost spirit, echoing and re-echoing around the hills," while the blood-curdling wails of common loons have been sometimes mistaken for the howls of gray wolves. According to the Norse people, if loons flew overhead calling, they were following souls to heaven.

COMMON LOON

CLASS	**Aves**
ORDER	**Gaviiformes**
FAMILY	**Gaviidae**
GENUS AND SPECIES	***Gavia immer***

ALTERNATIVE NAME
Great northern diver (Britain only)

WEIGHT
8–10 lb. (3.6–4.5 kg)

LENGTH
Head to tail: 28–36 in. (70–90 cm); wingspan: 4¼–5 ft. (1.3–1.5 m)

DISTINCTIVE FEATURES
Streamlined body; long, spearlike bill; partially webbed feet set far back on body. Summer: black bill; bright red eye; glossy black head and upperparts checkered with white. Winter: grayish bill; grayish brown head and upperparts; white throat and neck.

DIET
Mainly fish; also crustaceans and mollusks

BREEDING
Age at first breeding: 2 years; breeding season: eggs laid late May–June; number of eggs: 2; incubation period: 24–25 days; fledging period: probably 70–77 days; breeding interval: 1 year

LIFE SPAN
Up to 20 years or more

HABITAT
Summer: pools and lakes of subarctic zone. Winter: mainly rocky coastlines.

DISTRIBUTION
Summer: Canada and northernmost U.S.; parts of Greenland; Iceland. Winter: Pacific and Atlantic coasts of North and Central America; coasts of northwestern Europe.

STATUS
Locally common

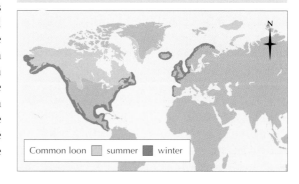

Common loon ▢ summer ▪ winter

Waterside nests

Loons form a strong pair-bond and pairs probably stay together for life, returning to the same nest site every year. Courtship takes place on the water, with the pair of loons chasing each other, either splashing, half-flying across the surface or swimming with the body partly submerged and the neck held out stiffly. Mating takes place on the nest or in the water, both birds submerging for part of the time.

The nest is usually no more than a depression in a hummock near the water's edge, but the common loon sometimes builds a heap of decaying vegetation. A slipway may be formed as the loons shuffle to and fro.

Incubation of the two or rarely three eggs is shared by the parents. The black fluffy chicks emerge after about 24–25 days and soon take to the water. At first they cannot dive and can swim only weakly; the ripples from the slightest breeze keep them pinned against the bank. They often ride on their parents' backs or retire to the nest when tired, but soon they are able to dive well, disappearing underwater for 6–7 seconds if danger threatens.

The chicks grow rapidly and their diving ability improves apace. Their parents continue to feed them for up to 6 weeks in the case of the red-throated loon, or up to 10–11 weeks in the common and yellow-billed loons. Young loons do not reach sexual maturity for 2–3 years.

Dangers to the chicks

If a loon is disturbed on the nest it freezes with neck stretched low, and if further disturbed it slips quietly off the nest and into the water, submerging and surfacing some distance away. Unless a predator sees the loon leave the nest it will not know a nest is there. Foxes, mink, otters, crows, jaegers and gulls are the main enemies of loons. Another danger is sudden changes in water levels due to heavy rain, which can flood the waterside nests.

Prophesying weather

In the Shetland Islands, north of Scotland, red-throated loons are traditionally called rain geese. They have long been said to foretell rain when their cry, which sounds rather like "we're all wet, worse weather," rings out over the lakes and barren moorland. There are also various local sayings connecting the movements of loons and the weather. For example, one says that when the loons fly inland to their breeding lakes up in the hills in late spring it is time to put the fishing boats to sea.

Red-throated loons breed on smaller pools than the other three species, some no more than 65 feet (20 m) long. They often commute to larger waters nearby to fish.

LORIKEET

LORIKEETS ARE COLORFUL little parrots from eastern Indonesia, New Guinea, Polynesia and Australia. They are predominantly green, usually an olive green, with small bills and short to moderately long tails. There are 35 species of lorikeets, which range in size from 6–15 inches (15–38 cm). All of them have areas of other colors, particularly reds, blues and yellows, but also purple, brown and black, in patches on the head, neck, breast and belly.

Together with the lories, the lorikeets or honey parrots make up the subfamily Loriinae of the parrot family Psittacidae. There are seven genera of lorikeets and many of the species are divided into subspecies, some with their own common name.

All the colors of the rainbow

One of the smallest species is the little lorikeet, *Glossopsitta pusilla*, at just 6–6⅓ inches (15–16 cm) long. It is greenish, with a black bill and brilliant crimson face, and is hard to see as it clambers mouselike among foliage and blooms. The purple-crowned or blue lorikeet, *G. porphyrocephala*, is fractionally larger, with a purple crown, patches of red and yellow on the head and a pale blue throat and breast changing to yellow in the back. The rainbow lorikeet, *Trichoglossus haematodus*, 10–12 inches (25–30 cm) long,

A rainbow lorikeet feeding on the nectar of a hibiscus flower. Lorikeets have a long tuft-tipped tongue for mopping up nectar and pollen, which accounts for up to 90 percent of their food intake.

just fails to have in its plumage literally all the colors of the rainbow. Its back, wings and the upperside of the tail are the usual green, but elsewhere the plumage is most variegated and is set off by a bright red bill. The rainbow lorikeet is divided into about 20 subspecies, which range over the islands of southeastern Indonesia and Polynesia as well as throughout the coastal regions of northern and eastern Australia. The scaly-breasted lorikeet, *T. chlorolepidotus*, is so named because its breast feathers look like overlapping scales. It displays little variation from the typical greenish color, until it raises its wings and shows their red and yellow undersides.

Colorful nomads

Lorikeets are swift in flight and usually move about in flocks. They are to a large extent nomadic, following the seasonal blossoming of eucalyptus and other flowering trees. As the lorikeets climb about the trees, they keep up a high-pitched chattering and screeching, and the combination of bright blossoms with brilliant birds intermingling is wonderful to the human eye. As suddenly as a flock descends on a flowering tree so it will take flight again, still chattering, the birds' combined wings creating a loud rushing sound. Many local ornithologists would find it difficult to imagine the Australian landscape lacking these happy honey-lovers that go shouting and screeching from district to district throughout the year, following the flow of gum tree blossoms.

Lorikeets generally take little notice of people. On the contrary, some species seem to be especially tame, and this led to the Griffiths sanctuary in Queensland, Australia, where large numbers of rainbow and scaly-breasted lorikeets come each day to be fed by hand. The sanctuary is a popular visiting place, and the lorikeets perch freely on the visitors' heads and hands. Elsewhere in Australia people have similarly attracted lorikeets by putting out sugar water.

Important pollinators

Lorikeets feed mainly on flower nectar, and they play a major part in the pollination of some of the trees they visit. Other nectar-eating birds have thin tubular bills for siphoning the sweet liquids. However, lorikeets crush the blossoms with their bills and then lap up the juices with their tongues as well as inserting the tongue skillfully into the flowers to lap up nectar with its brushlike tip. The ground beneath a tree may be littered with fallen blossoms broken off by the

RAINBOW LORIKEET

CLASS	**Aves**
ORDER	**Psittaciformes**
FAMILY	**Psittacidae**
SUBFAMILY	**Loriinae**
GENUS AND SPECIES	***Trichoglossus haematodus***

ALTERNATIVE NAMES
Rainbow lory; green-naped lorikeet; red-breasted lorikeet; red-collared lorikeet; blue-bellied lorikeet, coconut lorikeet

WEIGHT
2⅔–5⅔ oz. (75–160 g)

LENGTH
Head to tail: 10–12 in. (25–30 cm)

DISTINCTIVE FEATURES
Small, slender parrot with very long tail; coloration varies according to subspecies. *T. h. moluccanus* (eastern and southern Australia): blue head; red bill; yellowish green collar; green upperparts and tail; reddish breast; blue belly.

DIET
Almost entirely flower nectar and pollen from trees and shrubs; some fruits

BREEDING
Breeding season: all year (New Guinea and northern Australia), eggs laid July–January (southeastern Australia); number of eggs: 1 to 3; incubation period: 25 days; fledging period: 49–56 days; breeding interval: 1 year

LIFE SPAN
Up to about 10 years

HABITAT
All types of lowland wooded country

DISTRIBUTION
Eastern Indonesia; New Guinea; northern and eastern Australia; southwestern Pacific

STATUS
Abundant in much of range

Rainbow lorikeet

lorikeets in their eagerness to get at the nectar. They also eat buds and fruits, and flocks of lorikeets visiting orchards can be a pest.

Hole nesters

The simple nest lacks any soft lining materials and is located in a hole in a tree trunk or in the cavity at the end of a broken branch, fairly high above ground. The breeding season varies according to species and region, but is often year-round in Indonesia, New Guinea and northern Australia and from July to January in southeastern Australia. A typical clutch is of one to three white eggs, which are incubated by the female alone for about 3–4 weeks. The chicks soon become covered with a gray down and leave the nest 7–8 weeks after hatching.

Among the chief predators of lorikeets are birds of prey. A flock of lorikeets passing overhead dives for cover if it hears the call of the Australian or brown goshawk, *Accipiter fasciatus*.

The olive-headed or perfect lorikeet, Trichoglossus euteles, is natve to hillside forest on several islands of the Lesser Sundas group.

LORIS

shank is a network of blood vessels called the rete mirabile. This slows down blood flow and allows the animal to cling onto a branch for long periods at a time, such as when it is sleeping. Lorises have prominent ears and, being nocturnal, very large eyes. In common with the lemurs, they have a claw on the second toe and comblike front teeth, both of which are used in grooming.

Slow-moving trio

There are three species of lorises. The slender loris, *Loris tardigradus*, of southern India and Sri Lanka, is about 10–15 inches (25–38 cm) long and is grayish brown in color with black eye rings. It has a rounded head, pointed muzzle and long, slender limbs, the forelimbs being almost sticklike.

The slow loris, *Nycticebus coucang*, is similar in size to the slender loris but much plumper. Its limbs are also somewhat shorter and stouter. The muzzle is rounded and it is more brown than gray in color. It has a brown or black back stripe that forks on the crown of its head, sending a branch to each ear and eye, joining with the eye rings. The forepart of the back, especially surrounding the dorsal stripe, tends to have long, frosted tips to the hairs, making it look ashy or silvery white.

The pygmy slow loris, *N. pygmaeus*, is also known as the lesser slow loris. It has less woolly hair than the slow loris and is only 7–10½ inches (18–26 cm) long. All three lorises are tailless. The slow loris is found from Bengal, India, and Indochina southward to Java and Borneo, whereas the pygmy slow loris is restricted to southern China, Vietnam, Cambodia and Laos. Both species of slow lorises are often found in the same forest, but they avoid competing for food and do not interbreed.

Studies have shown that the two loris genera are not as different as was formerly supposed. In many of its skull and teeth characteristics, the pygmy slow loris partially bridges the gap between the other two species. In common with the slender loris, it has a long snout and partially obliterated markings. However, it is also plump, its body shape being more like that of the slow loris.

A slow loris eating a stick insect. The loris also hunts small birds, reptiles and mammals, and feeds on fruits and fresh leaves.

LORISES ARE SLOW-MOVING primates of Asia, closely related to the potto, *Perodicticus potto*, and the golden potto or angwantibo, *Arctocebus calabarensis*. They have thick coats and broad, grasping hands and feet. Each hand has an enlarged, opposable thumb and a much reduced index finger, although not as reduced as in the pottos of Africa. In each forearm and

LORISES

CLASS	**Mammalia**
ORDER	**Primates**
FAMILY	**Lorisidae**

GENUS AND SPECIES **Slender loris, *Loris tardigradus*; slow loris, *Nycticebus coucang*; pygmy slow loris, *N. pygmaeus***

ALTERNATIVE NAME
Pygmy slow loris: lesser slow loris

WEIGHT
**Slow loris: ⅖–4½ lb. (0.4–2 kg).
Pygmy slow loris: 3–12½ oz. (85–350 g).**

LENGTH
Slender loris and slow loris, head and body: 10–15 in. (25–38 cm). Pygmy slow loris, head and body: 7–10½ in. (18–26 cm).

DISTINCTIVE FEATURES
Huge eyes; prominent ears; grasping hands and feet with opposable thumbs and large toes; thick gray or brown coat

DIET
Large invertebrates; small reptiles, birds and mammals; fruits and fresh leaves

BREEDING
Age at first breeding: 10 months (female), 18 months (male); breeding season: all year; number of young: 1 or 2; gestation period: 165–200 days; breeding interval: 1 or 2 litters per year

LIFE SPAN
Up to 15–25 years

HABITAT
Forests, dry woodland and bamboo groves

DISTRIBUTION
Slender loris: southern India and Sri Lanka. Slow lorises: Southeast Asia.

STATUS
Slender loris and pygmy slow loris: vulnerable and declining. Slow loris: uncommon.

Lorises

Creeping up on prey

Lorises are arboreal (tree-living) and move about the trees in a slow, deliberate manner, with only one foot being moved at a time. Slow lorises tend to be slower moving than slender lorises, hence the name. The painstaking hand-over-hand movements are thought to help lorises creep up, unobserved, on their prey. When it is within reach, the loris will swoop down on its victim, grabbing the animal with its hands. Lorises often hang from branches by their feet, leaving their hands free to grasp food. In addition to large invertebrates and small birds, reptiles and mammals, lorises also feed on fruits and leaves. The slow loris and the pygmy slow loris are more likely to feed on plants and vegetation. The slender loris is more or less carnivorous.

The slender loris (above) differs from both slow loris species in having long, slender limbs. All lorises have huge eyes.

Day sleepers

The slender loris is found in montane forest, tropical rain forest, swamp forest and dry woodland. The slow lorises, on the other hand, are restricted to tropical rain forest and bamboo groves. All three species are solitary, sleeping by day and coming out to hunt at night. They roll themselves into a ball to sleep, grasping a branch using their hands and feet and tucking their heads down between their limbs.

In captivity, the slender loris is generally rather bad-tempered. The slow loris is also bad-tempered at first but more often becomes tame. When disturbed or threatened, lorises growl and make a high-pitched chatter and try to bite. The firmness with which they can grip tree branches is instinctive, being highly developed at birth.

A female slow loris and her young. The baby is born well-developed and is able to cling to its mother from birth.

Clinging to mother

Lorises breed throughout the year in the wild. The slender loris is said to come into season twice a year, with a 5½-month interval. Its gestation period is 165–170 days. Gestation is a little longer in the slow loris, between 185 and 200 days. The female slow loris comes in season every 42 days or so.

Lorises give birth to one, sometimes two, young. The baby is born well-developed and clings to the mother's fur. Sometimes the mother places it on a branch while she goes to forage. The baby clings tightly to the branch until she returns. The young loris is weaned at between 5 and 7 months, but might stay with its mother for up to a year. Female lorises reach sexual maturity at 10 months, males at 18 months.

Hunted for their eyes

Both the slender loris and the pygmy slow loris are listed as being vulnerable. This is not only a result of habitat loss, but also because of humans taking them as pets or using their body parts in traditional medicine. The lorises' huge eyes have been their downfall in this respect. These slow-moving, ghostly inhabitants of the tropical forests, stealthily crawling along branches, are the source of a great many superstitions. The slender loris, for example, is captured and sold live in southern India because its large eyes are sometimes believed to be love charms. Some doctors in Tamil Nadu, southeastern India, use its eyes in medicines they prescribe for eye diseases. Similar tales have evolved concerning slow lorises.

Whistle down the wind

Another belief of past times arose because of the single-note whistle slow lorises make at night. Nobody knows the purpose of this call. Since it is made with increased frequency by the female as she comes into season, it may be a mating call. Chinese sailors used to take these animals with them to sea. There they would listen for the slow lorises' whistle, which, they believed, indicated the approach of a wind.

LORY

EIGHTEEN COLORFUL species of parrots are called lories, and there are also numerous subspecies. Lories are placed in five genera: *Lorius* and *Eos* (both with six species), *Chalcopsitta* (four species) and *Phigys* and *Pseudeos* (each with a single species). Lories are closely related to the lorikeets, a group of parrots in the same subfamily, but lories are bigger and have short tails, rounded or square at the end, instead of the long pointed tails of the lorikeets.

Lories live on the islands of southern Indonesia and the southwestern Pacific, notably New Guinea.

Plumage dominated by red

The color of lories is predominantly red, supplemented with smaller areas of yellow, purple and green. By contrast, lorikeets are mainly green with yellow and red patches.

The yellow-backed lory, *Lorius garrula*, is brilliant red with a splash of yellow on the back, green wings and a horn-colored or orange bill. Its eyes are bright amber and its feet blackish gray. It is about 11 inches (28 cm) long and occurs in the Moluccas. The purple-naped or purple-capped lory, *L. domicella*, also of the Moluccas and other nearby islands, is roughly the same size. It is bright scarlet with green wings, blue shoulders and legs and a blackish purple crown. A broad yellow band tinged with red crosses the upperpart of the breast, and the tail feathers show a dark band just behind the yellow tips. The bill is orange yellow. Other species in the genus *Lorius* look quite similar. The dusky lory, *Pseudeos fuscata*, has a pinkish red or pinkish orange plumage crisscrossed with a complex pattern of bluish purple bars and patches.

A taste for sweets

Lories are very like lorikeets in their habits. Both of these groups are brush-tongued, which means they have slender tongues with a brush of long fleshy filaments, like thick hairs, at the tip. Most other parrots have relatively short, thick and fleshy tongues. The lories' specialized tongue shape is an adaptation to their main diet of flower nectar, pollen and blossoms, although they also eat seeds and the pulp of certain fruits, such as figs and the tiny fruits of coconut trees.

Judging from some lories kept in captivity, soft-bodied insects are sometimes eaten. In one instance in which nestling lories were hand-fed, they took mixtures of honey, apple puree and various baby foods, but when offered mealworms, showed a strong preference for them.

Breeding

Much of our knowledge of the lories' breeding habits has come from observations of captive birds. Lories can be difficult to observe in their forest habitat and there have been few research programs dedicated to them.

The yellow-backed lory has bred relatively frequently in aviaries. Two rounded white eggs are laid and incubated for slightly less than a month. In one case the male shared the incubation; in others it is not clear whether he did or not. In all cases the eggs were laid in a hollow log or an equivalent box, with peat or sawdust on the floor, and with an opening so small that it was difficult to see what was going on inside. In the wild lories nest inside holes in tree trunks, usually natural cavities caused by decay or a branch falling away.

Lories need to be acrobatic to reach their favorite foods, which include flower nectar, pollen and blossoms. Pictured is a female dusky lory of New Guinea.

Lories live in southern Indonesia and the southwestern Pacific, where deforestation and trapping is a growing problem. Threatened species include the purple-naped lory (above).

The nestlings first grow a covering of gray down; then in 3 weeks feathers begin to sprout and the eyes open. The parents feed them by regurgitation. At about 1 month nestling yellow-backed lories have green wings, but without flight feathers and the yellow on the back. The red does not appear until later. The bill is black, as is the eye. Final fledging comes about 3 months from hatching.

Mixed history

Some species of lories have long been favorite cage birds, not only because of their colors but also for their ability to mimic human speech, the purple-naped lory being an example. Apart from studies of aviary birds, information about lories has been obtained from museum specimens. These collections largely date from the 19th century and were created by planters and merchants, and by explorers such as Alfred Russell Wallace. Today several lories are at risk because of habitat destruction and trapping for the pet trade. The red and blue lory, *Eos histrio*, is endangered and three other species are considered vulnerable.

PURPLE-NAPED LORY

CLASS **Aves**

ORDER **Psittaciformes**

FAMILY **Psittacidae**

SUBFAMILY **Loriinae**

GENUS AND SPECIES *Lorius domicella*

ALTERNATIVE NAME
Purple-capped lory

WEIGHT
5¼–7¾ oz. (150–220 g)

LENGTH
Head to tail: about 11 in. (28 cm)

DISTINCTIVE FEATURES
Medium-sized parrot with brushlike tongue and square-ended tail; orange bill; blackish purple crown; rest of head, underparts and tail bright scarlet; broad yellow breast band; green wings; blue legs and shoulders

DIET
Seeds of rattan palm and other plants; also fruits and flower blossom, pollen and nectar

BREEDING
Poorly known (information relates to birds in captivity). Number of eggs: 2; incubation period: about 24 days; fledging period: about 90 days.

LIFE SPAN
Not known

HABITAT
Hill forest at altitude of 1,300–3,600 ft. (400–1,100 m)

DISTRIBUTION
Indonesian islands of Seram and Ambon (part of the southern Moluccas)

STATUS
Vulnerable; uncommon to rare on Seram, extremely rare or extinct on Ambon

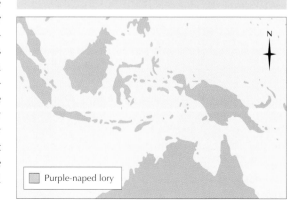

Purple-naped lory

LOVEBIRD

LOVEBIRDS ARE TINY African parrots named for their sociable habits. Pairs spend much of their time huddling together, bill to bill. They mate for life, like many other parrots, and the French and Spanish names for these birds mean "inseparable." Lovebirds are popular cage birds because of their small size, brilliant plumage and "loving" habits.

Sparrow-sized parrots

The nine species of lovebirds are about 5–6 inches (13–15 cm) long, parrotlike in form and generally green in plumage. The back and wings are usually darker green than the underparts, and most species have a gray or blue rump. Five species have patches of red on the head: the black-winged or Abyssinian lovebird (*Agapornis taranta*), red-headed lovebird (*A. pullarius*), rosy- or peach-faced lovebird (*A. roseicollis*), Fischer's lovebird (*A. fischeri*) and Nyasa or Lilian's lovebird (*A. lilianae*). Two species have variable blackish patches on the head: the black-cheeked lovebird (*A. nigigenis*) and masked or yellow-collared lovebird (*A. personatus*). The Madagascar or gray-headed lovebird (*A. canus*) has a pale gray head and breast, while the black-collared or Swindern's lovebird (*A. swindernianus*) has a thin black collar on the back of the head.

The precise taxonomic relationships of the various species of lovebirds have been much debated by ornithologists and cage bird enthusiasts alike. Most authorities recognize the nine species described above. The closest relatives of the lovebirds are the hanging parrots (genus *Loriculus*) of Asia.

Found in savanna and bush

Except for the Madagascar lovebird, all lovebirds live on the mainland of Africa south of the Sahara Desert. The red-headed and black-collared lovebirds are found in many parts of equatorial Africa and the rosy-faced lovebird occurs in southwestern Africa. The black-cheeked lovebird has a restricted range in Zambia and is now thought to be endangered.

Lovebirds are typically birds of arid, fairly open country with scattered trees or scrub. They prefer savanna and bushland and avoid thick forests, staying near the margins or in clearings. Lovebirds eat a variety of foods including seeds, fruits and nectar, although seeds are by far the most important. In places flocks of lovebirds do some damage to agricultural crops such as rice and maize; this is especially true of the rosy-faced lovebird.

Behavior

Lovebirds can be divided into two groups on account of their behavior. In the Madagascar, red-headed and black-winged lovebirds the sexes have different plumage and each pair nests on its own, whereas in the other species there is no difference in plumage between the sexes and group nesting is the rule.

Lovebirds also share certain habits with the hanging parrots. The latter are sometimes known as bat parakeets, and both names refer to their habit of roosting upside down. Hanging parrots hang from the branches of leafy trees in clusters and preen each other in this position after waking. The red-headed lovebird is the only lovebird known to behave in this remarkable way, but the Madagascar lovebird has been seen to bathe by hanging upside down in the rain with its wings and tail outspread.

Another strange trait some lovebirds share with hanging parrots is that of carrying nest material in their feathers. All lovebirds prepare

A pair of Fischer's lovebirds. Most parrots establish a lifelong bond with the same mate, but lovebirds are exceptional because they almost invariably roost in pairs side by side and spend hours preening one another.

MASKED LOVEBIRD

CLASS	**Aves**
ORDER	**Psittaciformes**
FAMILY	**Psittacidae**
GENUS AND SPECIES	***Agapornis personatus***

ALTERNATIVE NAME
Yellow-collared lovebird

WEIGHT
1½–1⅔ oz. (43–47 g)

LENGTH
Head to tail: 5–6 in. (13–15 cm)

DISTINCTIVE FEATURES
Tiny, compact parrot with short tail. Adult: bright red bill; blackish brown head; white ring around eye; yellow breast, extending onto nape of neck to create collar; green belly, upperparts and tail. Juvenile: duller plumage on head.

DIET
Seeds of various plants, especially grasses and millet

BREEDING
Age at first breeding: not known; breeding season: eggs laid March–April and June–August (dry months); number of eggs: 3 to 8; incubation period: about 23 days; fledging period: about 44 days; breeding interval: not known

LIFE SPAN
Probably up to 10 years

HABITAT
Well-wooded bushland; acacia thorn scrub at altitudes of 3,600–5,900 ft. (1,100–1,800 m)

DISTRIBUTION
Central Tanzania

STATUS
Locally common

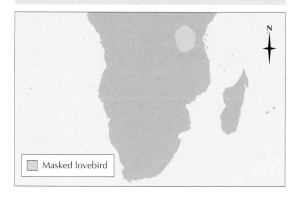

Masked lovebird

Lovebirds (masked or yellow-collared lovebird, above) eat the seeds of grasses, weeds and trees such as acacia. Because of this dry diet they must regularly visit waterholes to drink.

nest material by punching pieces out of leaves or tearing them into strips. The Madagascar, redheaded, black-winged and rosy-faced lovebirds carefully tuck the pieces of nest material into their feathers to carry them to their nest holes. The feathers are erected, half a dozen pieces of material are pushed in and the feathers are flattened to hold them in place. The five other species of lovebirds carry nest material in the normal way, in their bills.

Breeding

Lovebirds nest in holes excavated out of tree trunks or inside the earthy nests that some termites make on branches. Some species construct complicated nest chambers with entrance tunnels inside the nest holes.

At the start of courtship the male is wary of the female and alternately sidles toward her and then retreats, twittering and bobbing his head. In some species he feeds her, while in others the female feeds the male. The eggs are incubated for just over 3 weeks, and until the chicks fly, at 6–7 weeks old, they are fed by both parents. Toward the end of this period the male does most of the feeding. If disturbed, lovebirds that nest in colonies mob the intruding predator, screaming at it and flapping their wings.

LUGWORM

LUGWORMS ARE LARGE WORMS living in sand on the seabed. Their coiled castings are a familiar sight on a beach at low tide, but the animals themselves are not seen except by those who dig the worms out of the sand, normally to use as fish bait. There are about 28 species of lugworms belonging to four different genera.

When fully grown the lugworm, *Arenicola marina*, of the coasts of Europe, is up to 9 inches (23 cm) long and about ⅜ inch (1 cm) in diameter, although this depends on its state of contraction during movement. Other species on the North American coasts, such as *A. cristata*, range from 3 to 12 inches (7.5–30 cm) in length. The body is, like that of earthworms, ringed or segmented. It is usually separated into two or three regions, for example, head, body and tail. Lugworms are generally fat, and greenish, reddish or yellow black in color. Bristles and gills may be present in different positions depending on species. For example, the body of *A. marina* is soft, with a harder tail that has no bristles or gills. Another species, *A. ecaudata*, has no tail and bristles all along its body. In all lugworms there is a well-developed system of blood vessels and the blood is red, because it is rich in the oxygen-carrying pigment, hemoglobin.

Life in a burrow

Lugworms live in U-shaped burrows in the sand. The "U" is made up of an L-shaped gallery lined with mucus, from the toe of which a vertical, unlined shaft runs up to the surface. This is the head shaft. At the surface the head shaft is marked by a small saucer-shaped depression. The tail shaft, 2–3 inches (5–7.5 cm) from the head shaft, is marked by a much coiled casting of sand at the surface.

The lugworm lies in this burrow with its head at the base of the head shaft, swallowing sand from time to time. This makes the column of sand drop slightly, so there is a periodic sinking of the sand in the saucer-shaped depression. When it first digs its burrow, the lugworm softens the sand in the head shaft by pushing its head up into it with a piston action. After that the sand is kept loose by the worm irrigating its burrow; a current of water is driven through the burrow from the hind end by waves of contraction passing along the body.

Eating sand

The lugworm can move backward and forward in its burrow by waves of contraction and expansion of the body, using the bristles on the middle part of the body to grip the sides. It moves toward the head shaft to swallow sand and later moves backward so its rear end goes up to the top of the tail shaft in order to pass the indigestible sand out at the surface as a long thin cylinder. As the sand and sediment passes through the lugworm's stomach and intestine, small particles of dead plant and animal matter in it are digested.

The feeding action of the worm is via the front end of the throat, which is pushed out through the mouth as a swollen, oval proboscis (an extensible tube). The proboscis gives out a sticky secretion to which sand and particles of food adhere. It is then pulled in again and the material sticking to it is swallowed. In some species the action of swallowing takes place every 5 seconds and after 8 to 15 swallows the lugworm rests for a few minutes. It has been estimated that it takes 15–30 minutes for the sand to pass through the body, although other studies suggest about an hour. Then the lugworm moves backward through the burrow and ejects this as the cylindrical castings familiar on beaches.

Depressions and casts (piles of excavated sand) from three burrows belonging to lugworms of the species Arenicola marina. *In each burrow, a depression marks the head end and a cast marks the tail end.*

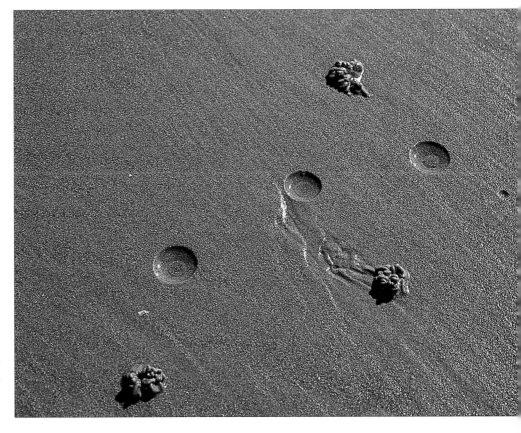

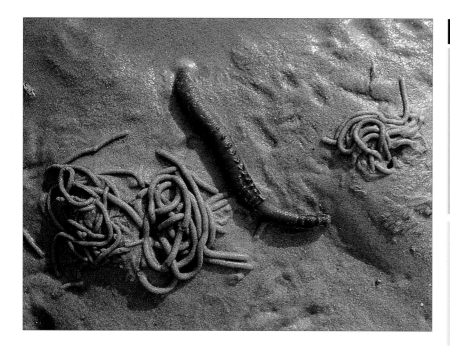

Lugworms seldom leave the safety of the sand except periodically to move burrows.

LUGWORMS

PHYLUM **Annelida**

CLASS **Polychaeta**

FAMILY **Arenicolidae**

GENUS **Abarenicola (16 species), Arenicola (9 species), Arenicolides (2 species), Branchiomaldane (1 species)**

SPECIES **About 28**

ALTERNATIVE NAME
Lobworm

LENGTH
Total length: usually up to 8 in. (20 cm), some species reach 12 in. (30 cm); diameter: ⅜–¾ in. (1–2 cm)

DISTINCTIVE FEATURES
Cylindrical, ringed worm, usually divided into 2 or 3 regions (for example, head, body and tail); bristles and gills present in various positions in some species; red, yellow or greenish black in color

DIET
Dead animal and plant matter taken from sand and sediment

BREEDING
Age at first breeding: 2–3 years, according to species; breeding season: September–October (*A. marina*)

LIFE SPAN
More than 3 years in some species

HABITAT
Coastal sand; found inside U-shaped burrows

DISTRIBUTION
Virtually worldwide

STATUS
Superabundant

Burrowing babies

Lugworms seldom leave the sand, and it is possible for them to remain in the same burrow for weeks on end. However, they sometimes move burrows every few days, leaving the old burrow and reentering the sand elsewhere.

Lugworms do not leave their burrows for breeding. Instead, they release their eggs and sperms into the water at the same time so that fertilization takes place. Breeding occurs around September or October in *A. marina*. The eggs are enclosed in tongue-shaped masses of jelly and each mass is anchored at one end. The larvae hatching from the eggs feed on the jelly and eventually break out when they have grown to a dozen segments and are beginning to look like their parents. They then burrow into the sand, usually higher up the beach than the adults, and gradually move down the beach as they get older.

Safe underground

Lugworms often fall prey to bottom-feeding animals and are sometimes eaten by seabirds probing the sand. They are most vulnerable in the young stages, when they are finding their first burrow or when seeking a new burrow as they move down the beach. Also, the burrows of young lugworms are nearer the surface while those of adults may be as much as 1 foot (30 cm) deep, sometimes twice this. Burrowing gives protection not only from living predators but also from the elements, such as changes in temperature and violent wave action.

Shift-working worm

Throughout each day a lugworm alternates bursts of activity with periods of rest. Each cycle is made up of several phases associated with feeding, irrigation of the burrow for breathing, and defecation. The lugworm swings forward in its tubular burrow to feed and then moves backward to defecate and breathe. It then swings forward again, thus driving water up into the sand in the head shaft and at the same time drawing water in from the tail shaft to irrigate the burrow. It is a kind of pumping action that makes the functions of living possible for the worm. All the various movements in this cycle of activity arise spontaneously.

LUMPSUCKER

THE LUMPSUCKER HAS a stocky build, growing up to 2 feet (60 cm) in length and 21 pounds (9.5 kg) in weight. Its body is rounded and humped, ornamented with rows of tubercles (prominent bumps or nodules), and it has a large head. The male lumpsucker is smaller than the female, and is deep blue in color. Its fins are nearly transparent and tinged with red. The male has a reddish belly, especially in the breeding season. The female is greenish with dark bluish patches. There is a reddish tinge on her pectoral fins and her belly is yellowish. On the underside of both sexes is an elaborate and efficient sucker formed by the pelvic fins.

This article deals only with the largest species of lumpsucker, *Cyclopterus lumpus*, which is widely distributed in shallow seas on both sides of the North Atlantic, but there are other, smaller species. A second group of related fish, known as snailfish or sea snails, family Liparidae, comprises 19 genera and 195 species. These have flabby or jellylike bodies covered with small spines instead of tubercles. The common sea snail lives throughout the North Atlantic.

Lumpsuckers spend much of their life clinging to rocks or, in the smaller species, to less solid supports, such as seaweed.

Eight months' fasting

The lumpsucker feeds mainly on comb jellyfish, marine worms, crustaceans, fish eggs and young fish. During and after the breeding season the male takes no food, so there is a period of fasting from April to November. The fish's stomach is distended with water during this time.

Writing in *The Fishes of the British Isles* (1925), Dr. J. Travis Jenkins noted that the fish's stomach is inhabited by a fungoid growth that clothes the whole mucous membrane of the stomach walls. A large bacillus is also often present in the fish's stomach. These organisms are able to exist in the lumpsucker's stomach due to the enormous amount of water that it always contains. If it were not for this water, the organisms would be destroyed by the stomach's digestive fluid.

Male stands guard

The female lays from 100,000 to 400,000 bluish red to yellowish eggs on the shore between the mid-tide and low-water marks. This means that at every tide they are uncovered for a period of time. The eggs do not form a solid mass but are spread over a rock surface and guarded by the male. While they are covered with water he fans them with his pectoral fins to ensure that eggs in

In Europe the lumpsucker is often referred to as the sea hen or hen fish. These names refer to the male's habit of caring for the eggs until they hatch.

Normally blue in color, the male lumpsucker turns bright orange or reddish in the breeding season. So striking is the transformation that it would be easy to mistake him for a different species.

the center of a clump will be fully aerated. It is probable that the male eats or otherwise removes any infertile or diseased eggs. After 6–10 weeks the eggs hatch into tadpolelike larvae.

In contrast to the slow-moving adult fish, the lumpsucker larva is very active. During rest intervals it holds on to a support with its sucker and wraps its tail around its large head.

Vulnerable eggs and larvae

Lumpsucker eggs and larvae are highly vulnerable to predators. When the tide is out the eggs are eaten by gulls, shorebirds, crows, starlings and rats. When the tide is in they are eaten by a variety of fish. Once the surviving eggs have hatched the larvae face the same dangers. The male guarding the eggs may be attacked by crows, torn from his perch by spring storms and battered on the rocks, or cast well up the beach, where he is likely to die.

The male lumpsucker has no means of defending the eggs. His primary role is to ensure that they have a good supply of oxygen and that diseased eggs do not survive to affect healthy ones. In the execution of this act he is genetically conditioned to guard the eggs until they hatch.

In one study, recorded in *The Fishes of the British Isles*, biologists watched a particular male lumpsucker for several weeks. Each time the tide went out the observers noted the fish remained by his eggs. When they removed the lumpsucker and placed him a couple of meters from his eggs he immediately wriggled back to take up position by them again. When moved to a greater distance he returned once more and used his sucker to affix himself in his former position with his snout almost touching the nearest eggs. On another occasion stormy waves dislodged the eggs, flinging them far up the beach. When the storm subsided the observers saw male lumpsuckers moving about over the shore, apparently searching for the lost eggs.

LUMPSUCKER

CLASS	**Osteichthyes**
ORDER	**Scorpaeniformes**
FAMILY	**Cyclopteridae**
GENUS AND SPECIES	***Cyclopterus lumpus***

ALTERNATIVE NAME
Lumpfish

WEIGHT
Up to 21 lb. (9.5 kg)

LENGTH
Up to 2 ft. (60 cm); female larger than male

DISTINCTIVE FEATURES
Rounded body; rows of large, coarsely spined plates on sides; large sucker disk on underside; first dorsal fin reduced (adult) or forms high crest (young). Male: deep blue to blackish (most of year), orange or reddish (breeding season). Female: greenish overall with dark blue patches.

DIET
Comb jellyfish, small crustaceans, worms, small fish and fish eggs

BREEDING
Age at first breeding: usually 5 years; breeding season: February–August; number of eggs: 100,000 to 400,000; hatching period: 42–70 days

LIFE SPAN
Up to 14 years

HABITAT
Rocky bottoms and floating seaweed in shallow seas

DISTRIBUTION
Western Atlantic: Hudson Bay and Labrador south to New Jersey; rarely to Bermuda. Eastern Atlantic: Greenland, Iceland and Barents Sea south to Spain.

STATUS
Common

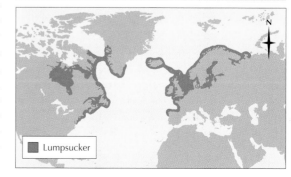

	Lumpsucker

LUNGFISH

THE EARLIEST FOSSILS OF lungfish go back over 350 million years. These and other fossils show that lungfish have, at some time or other, lived all over the world. Today only six species survive: one in Australia, four in tropical Africa and one in South America. As the name suggests, they have lungs for breathing air, so they can live in stagnant water. Some species can also survive the drying out of rivers.

The Australian lungfish, *Neoceratodus forsteri*, is the most primitive of the lungfish. Originally it was found only in the Burnett and Mary Rivers in Queensland but it has since been introduced into lakes and reservoirs elsewhere in that state. Sometimes called the Burnett salmon, it is a full-bodied fish. It may grow to as much as 6 feet (1.8 m) long and can weigh up to 100 pounds (45 kg), although it is usually only half this size. The body is covered with large scales, and the paired fins are flipperlike.

The largest of the African lungfish, *Protopterus aethiopicus*, measures 7 feet (2 m) but they are usually 2–3 feet (30–60 cm) in length. The South American lungfish, *Lepidosiren paradoxa*, grows up to about 4 feet (1.2 m). This and the four African species have eel-like bodies, with small scales embedded in the skin. The paired fins of the African lungfish are long, slender and flexible, whereas those of South American species are short and slender. The four African species range over a wide area of tropical Africa, while the South American lungfish is widely distributed over the Amazon River Basin.

Buried in mud

The African and South American species all have two lungs and must gulp air to live, regardless of the quality of the water. The South American lungfish tunnels into the mud if the stream, river or swamp it is living in dries out in summer. Two of the African species also burrow into the mud, but form cocoons by secreting slime that hardens.

This resting state that takes place in hot weather is called estivation. During estivation the fish continues to breathe, but at a reduced rate. The cocoon of the African lungfish, embedded in mud, has an opening to the surface. The fish lies doubled up with its head and tail at the lower end of this breathing tube. While estivating, the lungfish absorbs its own muscles for food, and one examined before and after 6 months resting had dropped from 13 to 10 ounces (370 to 280 g) in weight, while its length had decreased from 16 to 14½ inches (41 to 37 cm). Once out of estivation the fish more than made good these losses.

Apart from losing weight, another problem for the lungfish during estivation is the disposal of body waste. Its kidneys separate water from urea (the poisonous end product of food breakdown), the water going back into circulation while the urea is stored. In most vertebrates 10 parts per million of urea in the system would be fatal. A lungfish can survive 20,000 parts per million. As soon as estivation is over, the urea is discharged and the kidneys are cleared.

The fish leaves the mud or its cocoon when the rains return and the streams fill up. Usually the estivation lasts only a few months, but one West African lungfish, *Protopterus annectens*, was dug out while still in its cocoon, having been kept in a block of mud for over 4 years. It was immersed in water and, although emaciated, soon recovered and began to feed.

The Australian lungfish has only one lung and lives in water with plenty of oxygen, so it does not need to come to the surface to gulp air. It cannot survive drought as the other species can.

Carnivorous diet

The Australian lungfish eats both animal and plant food. Other species are mainly carnivorous, feeding on crustaceans, mollusks and small fish, although adults also eat some herbaceous stems. Live prey is caught by a powerful sucking action and then chewed before swallowing. Juveniles feed mainly on insect larvae and snails.

The South American lungfish has two lungs and is able to survive if the rivers, streams and lakes in which it lives temporarily dry up.

The largest African lungfish, Protopterus aethiopicus, reaches 7 feet (2 m) in length. It can survive drought by forming a cocoon and lying dormant until the rains return.

SOUTH AMERICAN LUNGFISH

CLASS	**Osteichthyes**
SUBCLASS	**Sarcopterygii**
ORDER	**Lepidosireniformes**
FAMILY	**Lepidosirenidae**
GENUS AND SPECIES	***Lepidosiren paradoxa***

LENGTH
Up to 4 ft. (1.2 m)

DISTINCTIVE FEATURES
Elongated, eel-like body; small scales embedded in skin; very short pectoral and pelvic fins; brown or olive green above, pale below

DIET
Adult: mainly bottom-living crustaceans, mollusks and small fish; also plant stems. Juvenile: insect larvae and snails.

BREEDING
Breeding season: beginning of rainy season; number of eggs: about 5,000; hatching period: not known; breeding interval: 1 year

LIFE SPAN
Up to several years

HABITAT
Stagnant waters with little or no current, including swamps, slow-moving rivers, streams and lakes; can survive without water by burying itself in mud

DISTRIBUTION
Brazil south to to Paraguay and Argentina

STATUS
Not threatened

South American lungfish

Male guards young

Reproduction takes place at the beginning of the rainy season. The primitive state of the Australian lungfish is also seen in its breeding habits. The female lays her eggs at random among water plants, and there is no parental care.

In the African and South American species, by contrast, the parents gather vegetal debris to create a nest in long burrows where female lays around 5,000 eggs. The male then tends the eggs, aerating them and chasing away anything that comes near. Once hatched, the larvae climb up the sides of the nest cavity and hang vertically with the head upward for 1–2 months. The larvae have four pairs of external gills and breathe water during the first weeks of life. These gills are gradually lost as the fish become aerial-breathing. The larva of the Australian lungfish does not have external gills.

Few predators in stagnant water

One of the main advantages for the lungfish of being able to live in stagnant waters is that they have few predators. That present-day lungfish have persisted more or less unchanged for millions of years suggests that this is the case. The main losses are among the young. Some of these may be eaten by the adults, especially in the African species, which are apt to take fish even of their own size and kind. Adult lungfish are also eaten by crocodiles and by fish eagles, which prey on the fish as they come to the surface to gulp air.

LUNGLESS SALAMANDER

THE NAME LUNGLESS salamander covers about 260 species of salamanders living in tropical and North America. They have neither lungs nor gills but breathe through their skin and the lining of the mouth. There are also seven European species, all of the genus *Speleomantes*, found in France, Italy and Sardinia.

Lungless salamanders range in size depending on species, from 2 to 8½ inches (5 to 22 cm). A few live permanently in water, but most are completely terrestrial. They are mainly somberly colored, black, gray or brown, but some have patches of red. The red-backed salamander, *Plethodon cinereus*, occurs in two color phases: red and gray, each with a black-and-white spotted belly. Varying proportions of red and gray individuals are found in any batch of its larvae.

Most lungless salamanders have the usual salamander shape: a long, rounded body and short legs, the front legs with four toes, the back legs with five. The tail is roughly the same length as the body. Exceptions include the four-toed salamander, *Hemidactylium scutatum*, which has four toes on each foot. The long-tailed salamander, *Eurycea longicauda*, is so called because its 7-inch (18-cm) tail dwarfs its 4-inch (10-cm) body. The California slender salamander, *Batra-*

choseps attenuatus, is snakelike, with vestigial (imperfectly developed) legs. It lies under fallen logs, coiled up tightly.

Some species are widespread. For example, the dusky salamander, *Desmognathus fuscus*, ranges over eastern North America from New Brunswick southward to Georgia and Alabama and westward to Oklahoma and Texas. Other species, such as the Ocoee salamander, *D. ocoee*, are very localized. It lives in damp crevices in rocks in Ocoee Gorge in southeastern Tennessee.

From deep wells to tall trees

Terrestrial lungless salamanders live mostly in damp places, under stones or logs, among moss, under leaf litter, near streams or seepages or even in surface burrows in damp soil. The shovel-nosed salamander, *Leurognathus marmoratus*, lives in mountain streams all its life, hiding under stones by day. Others live on land but go into water to escape predators. The pygmy salamander, *Desmognathus wrighti*, is only 2 inches (5 cm) long and lives in the mountains of Virginia and North Carolina. It can climb the rough bark of trees to a height of several feet. The arboreal salamander, *Aneides lugubris*, is able to climb trees to a height of 65 feet (20 m) and

The long-tailed salamander, so named because its tail is nearly twice the length of its body. Almost all lungless salamanders are found in tropical and North America.

sometimes makes its home in old birds' nests. One species of flat-headed salamander uses its webbed feet to walk over slippery rocks, using the tail for balance.

Several species live in caves or natural wells that go down to depths of up to 230 feet (70 m). All these species are blind. One species retains its larval gills throughout life, and one cave species spends its larval life in mountain streams but migrates to underground waters before metamorphosis, at which point it loses its sight.

Creeping, crawling food

All lungless salamanders eat small invertebrates. Those living in water feed mainly on aquatic insect larvae. Those on land hunt slugs, worms, wood lice and insect larvae. One group of lungless salamanders, genus *Plethodon*, are known as woodland salamanders. They live in rocky crevices or in holes underground and eat worms, beetles and ants. One species of woodland salamander, the slimy salamander, *P. glutinosus*, eats worms, hard-shelled beetles, ants and centipedes as well as shieldbugs, despite their strong odor. One of the European species catches its food with a sticky tongue.

Diverse breeding methods

Among the lungless salamanders there is as much diversity in breeding as in the way they live. Some lay their eggs in water and the larvae are fully aquatic. Others lay them on land, and among this second group are species in which the females curl themselves around their batches of 20 to 40 eggs as if incubating them. In a few species the female stays near her eggs until they hatch, but does not incubate them or give them

The eastern tiger salamander, Ambystoma tigrina, *is a large species that has both spotted and banded varieties.*

LUNGLESS SALAMANDERS

CLASS	**Amphibia**
ORDER	**Caudata**
FAMILY	**Plethodontidae**
GENUS	**About 30, including *Ambystoma, Aneides, Batrachoseps, Desmognathus, Ensatina, Eurycea, Gyrinophilus, Hemidactylium, Plethodon* and *Speleomantes***
SPECIES	**About 270 species**

LENGTH
2–8½ in. (5–22 cm)

DISTINCTIVE FEATURES
Adult (most species): long, rounded body with tail of similar length; short legs; 4 toes on front legs, 5 toes on back legs; mainly black, gray or brown in color with variety of markings. Larva: external gills.

DIET
Terrestrial species: mainly slugs, worms, ants, wood lice and insect larvae. Aquatic species: aquatic insect larvae.

BREEDING
Varies according to species. Breeding season: usually summer; number of eggs: 5 to 5,000; larval period: several weeks to several years.

LIFE SPAN
Not known

HABITAT
Most species: damp woodland, leaf litter, wet meadows, streamsides, seepages, rock piles and caves

DISTRIBUTION
Most species: tropical and North America; 7 species in France, Italy and Sardinia

STATUS
Critically endangered: 3 species; endangered: 4 species; vulnerable: about 25 species; many others locally common

Lungless salamanders ■ American ■ European

any special care. The woodland salamanders lay their eggs in patches of moss or under logs, and the larvae metamorphose before leaving the eggs.

Breeding in the dusky salamander
A typical species is the dusky salamander. The male deposits his sperm in a spermatophore, or capsule. He then rubs noses with the female. A gland on his chin gives out a scent that stimulates the female to pick up the spermatophore with her cloaca (a chamber into which the urinary, intestinal and generative canals feed). Her eggs are laid in clusters of about 25 in spring or early summer under logs or stones. Each egg is ⅙ inch (4 mm) in diameter, and the larvae on hatching are about ⅜ inch (9 mm) long. They have external gills and go into water. Here they live until the following spring, when metamorphosis takes place. The adults, 5 inches (13 cm) long, are dark brown or gray in color. When it first metamorphoses, the young salamander is brick red and light cream in patches. Later it takes on the colors of the fully grown adult.

Not so defenseless
Lungless salamanders, like other salamanders and newts, have few means of defence against predators. A possible exception is the arboreal salamander, which has fanglike teeth in the lower jaw. The slimy salamander gives out a sticky, glutinous secretion from its skin when handled, and this also possibly deters predators. Lungless salamanders are preyed upon by small snakes and frogs. Especially vulnerable are the larvae and the young salamanders.

The yellow-blotched salamander, *Ensatina croceator*, of California, has a curious behavior that may be defensive. It raises itself on the tips of its toes, rocks its body backward and forward, arches its tail and swings it from side to side. It also gives out a milky astringent fluid from the tail. More extraordinary, the yellow-blotched salamander is able to squeak like a mouse, which may also deter predators. Because it has neither lungs nor a voice box, it does this by contracting the throat, forcing air through the lips or nose.

Breath through the skin
The most distinctive common feature of these salamanders is that they lose their larval gills as they grow and never develop lungs. Instead, their skin has become the breathing organ, with the skin lining the mouth also acting the part of a lung. It does this by having a network of fine blood vessels in it, like the lining of a lung. The arboreal salamander has a similar network of fine blood vessels in the skin of its toes, and these may provide an additional breathing method.

Many salamanders (Jefferson salamander, Ambystoma jeffersonianum, above) lay their eggs in water. When the larvae hatch they have external gills and are fully aquatic. Later, as adults, they breathe through their skin.

LYNX

There are several forms of lynxes, which are usually separated into three distinct species: the Eurasian lynx (above), Spanish lynx and Canada lynx.

LYNXES ARE BOBTAILED members of the cat family. The bobcat, *Felis rufus*, of North America, stands somewhat apart from the others and is discussed elsewhere, as is the caracal or desert lynx, *F. caracal*.

The original animal to be given the name is now distinguished as the Eurasian lynx, *F. lynx*. It is 2½–4¼ feet (0.8–1.3 m) long, with a 4–10-inch (11–25-cm) tail. It weighs up to 84 pounds (38 kg) but the male averages 48 pounds (22 kg), while the female weighs about 40 pounds (18 kg). The Eurasian lynx has a relatively short body, tufted ears and cheek ruffs, powerful limbs and very broad feet. Its fur varies from a pale sandy gray to rusty red and white on the underparts. Its summer coat is thinner, often with black spots, while its winter coat is dense and soft, the spots being less prominent. The Eurasian species ranges through the wooded parts of Europe, and in Asia it extends eastward to the Pacific coast of Siberia and southward to the Himalayas.

Closely related and very similar in appearance is the Spanish lynx, *F. pardina*. It is generally smaller, 2½–3½ feet (0.8–1.1 m) in length with a 4½–12-inch (12–30-cm) tail. The male averages about 30 pounds (13 kg) in weight, while the female weighs about 20 pounds (9 kg). It has shorter and more heavily spotted fur than the Eurasian lynx. Also called the pardel, the Spanish lynx is now considered endangered and is found only in remote areas of Spain and Portugal.

The Canada lynx, *F. canadensis*, is similar in size to the Spanish lynx but has a shorter tail and longer hair. This species is often without spots, even in summer. It is widespread throughout North America.

The view of several zoologists is that lynx types all constitute one species, *F. lynx*, but for the moment we are following the accepted pattern.

Nocturnal hunters

Lynxes live in remote, wooded, mountainous areas. They are also found in forests, especially of pine, and in areas of thick scrub. They are solitary animals, hunting by night, using sight and smell. Lynxes are tireless walkers, following scent trails relentlessly for miles in pursuit of prey. They are also good climbers and sometimes lie out on tree branches, dropping onto their victims as they pass. Lynxes swim well, and their broad feet carry them easily over soft ground or snow.

The voice is a caterwauling similar to that of domestic tomcats but louder. Also like domestic cats, lynxes use their claws and teeth in fights. Within its home range a lynx buries its urine and feces, but near the boundary of this range both are deposited on prominent places, such as hillocks. These boundary marks are recognized by both the occupant and neighboring animals.

LYNXES

CLASS	**Mammalia**
ORDER	**Carnivora**
FAMILY	**Felidae**
GENUS AND SPECIES	**Canada lynx, *Felis canadensis;* Eurasian lynx, *F. lynx;* Spanish lynx, *F. pardina***

WEIGHT

**Canada lynx: 10–38 lb. (5–17 kg).
Eurasian lynx: 17–84 lb. (8–38 kg).
Spanish lynx: up to 28 lb. (13 kg).**

LENGTH

**Head and body: 2½–4¼ ft. (0.8–1.3 m);
tail: 2–12 in. (5–30 cm)**

DISTINCTIVE FEATURES

**Compact body; tufted ears; large, broad
feet; short bobtail. Eurasian and Spanish
lynxes: coat often spotted, especially in
summer. Canada lynx: usually spotless.**

DIET

**Small mammals such as rabbits, hares, pikas
and squirrels; also young deer, wild goats and
sheep, ducks, game birds, fish and insects**

BREEDING

**Age at first breeding: 1 year (female);
breeding season: late winter–early spring;
number of young: 1 to 4; gestation period:
63–74 days; breeding interval: 1 year**

LIFE SPAN

Up to 20 years

HABITAT

Remote, wooded mountains and thick scrub

DISTRIBUTION

**Canada lynx: northern North America.
Eurasian lynx: Scandinavia; Siberia south
to Central Asia; southeastern Europe.
Spanish lynx: parts of Spain and Portugal.**

STATUS

**Canada and Eurasian lynxes: uncommon;
rare in places. Spanish lynx: endangered.**

Lynxes ▢ Canada ▮ Spanish ▢ Eurasian

Northern subspecies tend to be larger and grayer than those found farther south. Pictured is a male Siberian lynx, Felis lynx wrangeli, a subspecies of the Eurasian lynx.

Instant death

Lynxes used to be numerous throughout Europe but are now scarce in most parts. Often they have been hunted and persecuted by humans because of their alleged raids on sheep, goats and other livestock. Some zoologists claim that lynxes are less interested in farmstock than in wild game. The natural food of the Eurasian lynx includes small mammals such as hares, pikas and rabbits, along with ducks and game birds. It will also take young deer and small, wild goats and sheep.

Prey are killed by a bite at the nape of the neck, which severs the spinal cord, or the lynx may use a two-way bite, into the shoulders and then into the nape. Death is instantaneous with both methods. Lynxes will also kill squirrels, foxes, badgers, fish, beetles, especially the wood-boring species, and many small rodents. They tend to kill small game such as rodents in summer, turning to larger game such as deer in winter. The Canada lynx has a similar diet. The snowshoe hare, *Lepus americanus*, is its main prey and the Canada lynx's population regularly increases and decreases in line with that of the snowshoe hare.

Slow developers

Mating takes place in late winter or early spring, the young being born after a gestation period of 67–74 days in the Eurasian lynx and 63–70 days in the Canada lynx. The litter is usually one to four kittens, born well-furred but blind. Their eyes open at 10 days and the kittens are weaned after about 3 months. The young remain with the mother for some time after weaning, until they are between 7 months and 1 year of age.

Although the kittens are somewhat advanced at birth, they are slow in developing. Even at 8 months or more they still have milk teeth and their claws are quite feeble. The young lynx must therefore feed on small rodents or food killed by its mother. Should it become separated from its mother when the first winter snows fall, its chances of survival are slim. The small rodents it needs to survive are able to live under 2 feet (60 cm) or more of snow. The female lynx matures at 1 year old, and the recorded life span is between 13 and 20 years in the wild. Lynxes are reproductively active for 10–14 years.

Lynxes have long been hunted for their fur, but populations are most at risk from destruction of their natural forest habitats.

Driven out of house and home

Humans are the only major threat to the lynxes, but the toll taken by human persecution and by habitat loss to agriculture and forestry has been a heavy one. The dense, soft winter coat has long been valued for garments and trimmings. In Canada, lynx fur was prominent in the transactions of the Hudson Bay Company.

More than hunting, the destruction or commercial management of forests in Europe and Canada has deprived the lynxes of their best and most natural habitat. In Sweden, for example, hunting and changes in the forest drove the Eurasian lynx northward during the 19th century. This meant less food and, more important, less chance of survival for the kittens, due to the longer and more rigorous winters in the higher northern latitudes. Finally, in 1928, the lynx was given legal protection in Sweden. Its numbers have since begun to rise and it is now ranging farther south again.

Although the Canada lynx remains widespread and there are some 100,000 of the Eurasian lynx in Russia and Asia, there are thought to be less than 200 of the latter species remaining in western Europe. It has recently been reintroduced to Austria, Switzerland and Slovenia. Most at risk, however, is the Spanish lynx, with only around 1,200 animals remaining in the wild.

LYREBIRD

IN 1798 THE EARLY EUROPEAN explorers of the mountain forests of eastern Australia discovered what they called a native pheasant. The name is confusing because the species they had found, *Menura novaehollandiae*, is not a true pheasant and because another Australian bird, the mallee fowl, *Leipoa ocellata*, is sometimes also known as the native pheasant. It was not until the 1820s that the name lyrebird came into use. Since then a second species has been discovered: the northern or Albert's lyrebird, *M. alberti*. The original species is known as the superb lyrebird.

Magnificent tails

The male superb lyrebird has a body approximately the size of a cockerel, with strong legs, feet and toes. His plumage is a warm brown tinged with red on the wings and gray on the underparts. His 2-foot (60-cm) long tail is made up of 16 feathers. The two outer feathers are broad and together resemble the U-shaped frame of a lyre, the harplike stringed instrument used in ancient Greece. There are also two slimmer curved feathers, known as guard plumes. The 12 remaining tail feathers are delicate and lacelike. The female superb lyrebird has a similar plumage but with an ordinary tail. Males do not start to grow the distinctive tail until they are 3 years old, and the full tail is acquired gradually over the following 6 years.

Legendary performance

In mountain forests where the rocky slopes, running down to fast flowing streams, are covered with large tree ferns, the lyrebirds act out their unique display. In the autumn each mature male lays claim to a territory of 3–6 acres (1.2–2.4 ha). With his strong legs and feet he scrapes together large mounds of earth and leaves on which to display. He may make a dozen of these in his territory. Having sung from the top of a log or from a low branch of a tree as a preliminary, he flies to the top of one of the mounds and begins to sing in a loud penetrating voice.

After a few minutes the lyrebird unfolds his tail feathers, which he has so far carried like a peacock's train. He raises these and swings them forward over his back, like a canopy, the two broad outer feathers that form the frame being swung out until they are at right angles to the body. Half-hidden under this shimmering canopy, he begins to dance, pouring out a torrent of bubbling notes. The song rises higher and higher and then suddenly stops. The tail is swung back into the normal position and the male lyrebird walks away, his display finished. Sometimes he shakes his tail feathers violently while they are spread over his back, making clicks and drumming noises at the same time.

One-chick family

The male lyrebird's elaborate display serves to maintain his territory and attract females. Mating is usually during May to July, when the female alone takes 3–4 weeks to build a large nest of sticks, lining it with moss. The male is polygamous, attempting to mate with several females, and gives no help with nest building. The nest is roofed and may be built on the ground, on a rocky ledge, in an old tree stump or in a high tree fork. In this the female lays one grayish purple egg, which she incubates for about 50 days.

The downy chick loses its black down 10 days after hatching and begins to grow feathers, staying in the nest for up to 7 weeks. During that time the female brings insects and snails to feed it. Adult lyrebirds feed by scratching the ground with their strong toes and claws, like domestic chickens, for insects and other invertebrates.

During his display the male superb lyrebird throws his tail over his body so that the two broad outer plumes are at right angles to him and the lacy central plumes cover him in a silvery shower.

Lyrebirds are shy by nature and tend to be unobtrusive in their dark and shady habitat. However, they have become tame in some forest parks where they have grown accustomed to the high numbers of visitors.

SUPERB LYREBIRD

CLASS	**Aves**
ORDER	**Passeriformes**
FAMILY	**Menuridae**
GENUS AND SPECIES	***Menura novaehollandiae***

ALTERNATIVE NAMES
Lyretail; native pheasant

LENGTH
Head to tail. Male: 3–3⅓ ft. (0.9–1 m); female: 2¾ ft. (0.85 m).

DISTINCTIVE FEATURES
Resembles a pheasant or (less so) chicken with especially strong legs, feet and toes. Male: rich brown upperparts with copper red tinge to wings; gray brown underparts; extremely long tail made of 12 lacy central feathers, 2 slim curved plumes and 2 broad curved outer plumes ending in a "club." Female and young male: shorter tail of 14 ordinary feathers.

DIET
Mainly beetle larvae, earthworms, spiders, millipedes and other soil-living invertebrates

BREEDING
Age at first breeding: at least 3 years; breeding season: eggs laid June–August; number of eggs: 1; incubation period: about 50 days; fledging period: about 46–49 days; breeding interval: 1 year

LIFE SPAN
Not known

HABITAT
Dense forests, especially in upland areas; often near to rocky gullies and ravines

DISTRIBUTION
Southeastern Australia; introduced to Tasmania

STATUS
Generally fairly common

The chief danger to lyrebirds is from snakes, large lizards and birds such as the kookaburra, *Dacelo gigas*, raiding their nests. Their only other enemy is humans, whose raids are now curbed because the two species are protected by law. However, in the early years of the 20th century lyrebird tails were still being sold on the streets of Sydney by the basketful.

Artistic error

The superb lyrebird appears on Australian seals and stamps, with its lyre-shaped tail held upright. This it never does except for a brief moment when swinging the tail from the train position to the canopy position over its back. Tradition dies hard, however, and even now artists are still drawing the lyrebird with his tail held in this unnatural position.

Mellow songs

The fame of the lyrebird rests not only on the males' remarkable tails but also on their voices. Their natural song is one of the most powerful of all songbirds. It is mellow and far-carrying. Lyrebirds are also superb mimics, incorporating the calls of many other birds in their song pattern. They can even mimic mechanical sounds.

Northern lyrebird Superb lyrebird

MACAQUE

THE MACAQUES ARE the most numerous and widespread of the Old World monkeys. They are found in the tropical forests of Southeast Asia, the scrublands of India, the snowy mountains of Tibet, the temperate forests of Japan and the Atlas Mountains of North Africa. They include forms with long tails and short tails and some species with no tail at all. Most are brown in color, but some are black. Macaques are larger and longer-faced than guenons, but smaller and shorter-faced than baboons. Two species, the Barbary ape, *Macaca sylvanus*, and the pig-tailed macaque, *M. nemestrina*, are discussed elsewhere.

Best known of the 20 species of macaques is the rhesus macaque, *M. mulatta*. It is an average sized macaque, 1⅔–2½ feet (50–75 cm) long with a 6–8-inch (15–20-cm) tail. It is brown with much brighter, more reddish hindquarters. Its face is pink. The rhesus macaque ranges from northern India west to Afghanistan, east to North Vietnam, south to central Myanmar (Burma) and Thailand, and north as far as the Yangtze River in China.

Long-tailed macaques

South of the Godavari River in India the bonnet macaque, *M. radiata*, replaces the rhesus, while the toque macaque, *M. sinica*, occurs in Sri Lanka. These two species are smaller than the rhesus macaque, under 1⅔ feet (20 cm) long, with unusually long tails for macaques at nearly 2 feet (60 cm). Both are entirely brown, with pink faces and a cap of radiating hairs on the head. In the bonnet macaque this cap is short, leaving the forehead bare. The crab-eating or long-tailed macaque, *M. fascicularis*, is found to the east, in southern Myanmar, Thailand, Peninsular Malaysia, Indonesia and the Philippines. It too is a small, long-tailed species, but has no cap. In the Himalayas, and more or less wherever the rhesus is found, but at higher altitudes, is the large, pale fawn Assam macaque, *M. assamensis*.

Stump tails or none at all

Most macaque species have very short tails, or none at all. The stump-tailed macaque, *M. arctoides*, has short, stout limbs and a bright red face. Old ones go bald. Also with short tails, the Tibetan macaque, *M. thibetana*, is larger, shaggy and bearded, with a brown face, while the Japanese macaque, *M. fuscata*, is almost as big, and also shaggy, but with a long, pink face.

The macaques of the Indonesian island of Sulawesi (formerly Celebes) are a world of diversity all by themselves. There is the jet-black,

long-faced Celebes black macaque, *M. nigra*, of the northern peninsula of Sulawesi. It has a backward-drooping crest of hair on its crown and pink, kidney-shaped ischial callosities (the patches of hard skin on the rump). There is also the brown, short-faced, crestless Moor macaque, *M. maura*, of southern Sulawesi. Many intermediate forms between these two species occur.

Large social groups

All macaques live in large social groups made up of all ages and both sexes. The exact size and composition of the group varies from place to place and from species to species. In sparse parts of forest and scrubland, even in towns and by railways or roadsides in India, rhesus macaque troops number 10 to 25 animals. In more fertile country and around Indian temples (where they are protected and even fed) troops of between 30 and 60 are common. Bonnet macaques live in somewhat larger groups, 25 to 30 in poor wild

Japanese macaques are among the largest of the 20 or so species of macaques. Now an endangered species, they range farther north than any other primates except humans.

Lion-tailed macaques,
Macaca silensu, *are the*
only macaques to have
a mane around the face.
They are native to hill
forests in southwestern
India, and have become
endangered.

MACAQUES

CLASS	Mammalia
ORDER	Primates
FAMILY	Cercopithecidae
GENUS	*Macaca*

SPECIES **20 including rhesus macaque,
M. mulatta; Japanese macaque, *M. fuscata*;
and toque macaque, *Macaca sinica***

WEIGHT
**Most species: up to 40 lb. (18 kg). Toque
macaque (smallest): 5½–13 lb. (2.5–6 kg).**

LENGTH
**Head and body: up to 2½ ft. (0.75 m);
tail: about 6 in. (15 cm). Toque macaque,
head and body: 1⅕–1⅔ ft. (35–50 cm).**

DISTINCTIVE FEATURES
**Long, naked face (pinkish in most species);
coat brownish (most species) or black
(several species); tail either absent or
stumplike (most species) or rather long
(a few species, including toque macaque)**

DIET
**Mainly fruits, leaves, shoots, bark, crops and
insects; also small vertebrates and scraps**

BREEDING
**Age at first breeding: 2–5 years (female),
4–6 years (male); breeding season: varies;
number of young: usually 1; gestation period:
150–180 days; breeding interval: 2 years**

LIFE SPAN
Average 25 years in captivity, less in the wild

HABITAT
**From high mountains to lowland rain forest,
bush, scrub, farmland and urban areas**

DISTRIBUTION
Southern and eastern Asia; Southeast Asia

STATUS
**Endangered: 4 species; vulnerable: 5 species;
most others declining but a few still common**

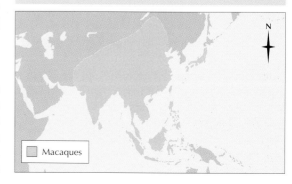

Macaques

environments and 50 to 60 in rich cultivated areas. Crab-eaters live in troops of approximately the same size or slightly larger. The largest groups of all used to be found among Japanese macaques, with an average troop size of 200 or so, although this species is now endangered.

When two macaque troops meet, there is almost always some kind of conflict. Normally, if the home ranges of two groups overlap, the timetables of daily movement are geared so they do not meet. If the smaller troop is unwary and fails to see the bigger troop coming, they are will often be attacked. Subordinate males generally begin the fight, which results in scratches and bruises, but rarely anything more serious.

Ordered societies

In rhesus, crab-eating and Assam macaque troops there are two or three females to every male. In bonnet troops the numbers are about equal. At the troop center is either a single dominant male or a coalition of two or more. Other males are ranked below the dominant ones. The coalition custom is also found in bonnets, but here the society is a little more relaxed and a subordinate male is allowed to sit in the center of the troop, provided he makes the necessary lip-smacking gestures and presents his hindquarters to the dominant males, as a sign of submission. As far as is known, crab-eaters have an even rank order with no coalition. In Japanese macaques, subordinate males are pushed right out to the periphery and may even live isolated, solitary lives.

Macaques share most of their range with langurs. There is no competition for food, macaques being omnivorous while langurs are leaf-eaters. The two types of monkeys coexist peacefully, but macaques are always dominant.

Mating and birth

Breeding is seasonal in areas where there is marked variation in food supply, year-round where there is not. In Jaipur, northern India, for example, there are two birth seasons for the rhesus macaque: March to May and September to October. Bonnet macaques in the same region show a birth peak in February and March. In Japanese macaques there is a peak in June and July. Sexual behavior also varies. In rhesus macaques a female in season forms a "consort pair" with a male. The pair move around and forage together and groom each other. The association lasts from a few hours to a few days.

Gestation is 150–180 days, the single young being born with its eyes closed, although they open within 2 hours. Hair on the newborn is mainly confined to the head and back and is usually darker than the adult's, so it is noticeable at a distance. Male macaques take quite an interest in the infants and tolerate their play. This fatherly attitude is especially marked among Japanese macaques. The bond between a mother and her offspring seems to continue throughout life, and the offspring of a mother that is of dominant rank inherit her high rank within the group.

Animal experiments

Macaques are preyed upon by leopards, smaller cats, wild dogs, eagles, pythons, crocodiles and even monitor lizards. However, it is humans that have been responsible for declines in many macaque populations. Over the years tens of thousands, mainly youngsters, have been exported to medical laboratories in Europe and the United States. Medical research depends to an great extent on experiments with live animals, and primates are the most valuable of all animals for this because they are so closely related to humans.

In purely scientific terms the great apes would be the ideal "blueprints" for human beings in such research. However, gorillas, chimpanzees and orangutans are rare and there

is also the ethical question of experimenting on these animals. The Old World monkeys are the next closest relative, and among these the macaques were found to be the most suitable. A notable vindication of this research was the development of the Salk vaccine against polio. Nonetheless, modern science is attempting to find alternatives to animal experimentation, and fewer macaques are now used in this way.

Some macaque species are still abundant but most have suffered declines, due mainly to their past and present use in medical research. Macaque numbers have also been reduced as a result of persecution and exportation for pets. Four species, the Moor, Celebes black, lion-tailed and Japanese macaques, are now endangered. Others, such as the stump-tailed, Assam, Tibetan and bonnet macaques, are vulnerable or threatened.

A long-tailed macaque, Sangeh Temple, Bali, Indonesia. Monkeys are such an important symbol in some Eastern religions that troops of macaques living around temples are protected and even fed.

MACAW

THE 18 SPECIES OF MACAWS include the largest members of the parrot family, Psittacidae. They live in tropical America, from central Mexico south through Central America to Paraguay and northern Argentina. Macaws have massive bills, the upper mandible (bill-half) being long and strongly hooked. In most species the skin on the cheeks and around the eyes is naked except for a few small feathers.

Spectacular parrots

Macaws are divided into six genera, of which *Ara* is the largest. The biggest species of macaw is the scarlet or red and yellow macaw, *A. macao*, found from southern Mexico to Bolivia and central Brazil. It grows to almost 3 feet (0.9 m) long, of which more than half is tail. Its plumage is mainly scarlet except for the yellow wing coverts and the blue of the flight feathers, lower back feathers and outer tail feathers. The red and green macaw, *A. chloroptera*, is very similar, but has green instead of yellow on the wings. The blue and yellow or blue and gold macaw, *A. ararauna*, which ranges from Panama south to Paraguay, is only slightly smaller. It is a rich blue on the crown, nape, back and wings and the upperside of the tail, and yellow on the underparts. There is a large black patch on the throat, the bill is black and the white sides of the face are marked with black wavy lines.

The military or great green macaw, *A. militaris*, is 32–33 inches (81–84 cm) long and ranges from Mexico to Brazil. It is predominantly green, shading to blue on the flight feathers, rump and tail coverts, with a crimson band on the forehead and red on the upperside of the tail. The hyacinth or hyacinthine macaw, *Anodorhynchus hyacinthinus*, is a large species that is cobalt blue all over except for a bright yellow eye ring and a yellow crescent of bare skin near the bill. Its range is limited to the rain forests of the Amazon Basin. The smaller species of macaws in the genera *Propyrrhura*, *Diopsittaca* and *Orthopsittaca* are generally green, with patches of other colors.

Macaws regularly visit soil banks to lick clay, which absorbs toxins in their diet. This site in Peru is used by scarlet macaws (distinguished by their yellow and blue wings) and red and green macaws.

SCARLET MACAW

CLASS	**Aves**
ORDER	**Psittaciformes**
FAMILY	**Psittacidae**
GENUS AND SPECIES	***Ara macao***

ALTERNATIVE NAME
Red and yellow macaw

WEIGHT
2–3⅓ lb. (0.9–1.5 kg)

LENGTH
Head to tail: 33–35 in. (84–89 cm)

DISTINCTIVE FEATURES
Huge size; strongly hooked bill; extremely long, graduated tail; bright red head and body; large area of bare white skin on face; wings mainly yellow and blue above and red below; red tail with blue tip

DIET
Seeds, nuts, fruits, flowers, leaves and bark

BREEDING
Age at first breeding: several years; breeding season: eggs laid March–April (Mexico and Nicaragua), October–April (Costa Rica, Venezuela and central Brazil); number of eggs: 2 to 4; incubation period: 24–28 days; fledging period: about 100 days; breeding interval: highly variable, but often 2 years or more

LIFE SPAN
Up to 40–50 years in captivity

HABITAT
Humid lowland forest, often near exposed riverbanks and clearings with large trees

DISTRIBUTION
Southern Mexico south to Panama; northern Colombia; Venezuela south to central Brazil

STATUS
Fairly common in undisturbed parts of South American range; rare in Central America

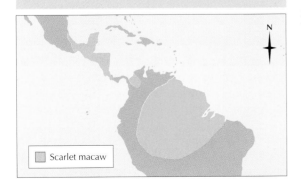

Scarlet macaw

A pair of hyacinth macaws at their nest hole. Macaws have a slow reproductive rate: they do not breed every year and take 14 weeks to raise their chicks to fledging age.

Commuting lifestyle

Macaws move about in screeching flocks except when breeding. Their day starts with a loud chorus of screeches as individual birds leave their roosts to gather in a tree. There they bask in the early morning sun before setting off to feed. As the midday heat builds up the macaws seek the shade, but when the sun's rays begin to weaken they become active again. At dusk they return to their assembly point, often a bare tree, before dispersing to roost.

Most macaws feed on seeds, nuts and fruits, the larger species cracking even hard-shelled nuts such as Brazil and palm nuts. They take foods from a host of tree species, including some, such as the soapbox tree, that are not targeted by any other vertebrates. Macaws eat unripe as well as ripe fruits; a tolerance of unripe fruit means they can feed from a tree before rival seed eaters such as bats, monkeys and other birds arrive on the scene. Some of the seeds and fruits eaten by macaws contain toxins, and this is probably the reason why they regularly visit streamside earth banks to lick the clay. It appears that the swallowed clay helps to absorb and filter out the harmful toxins.

Slow breeders

Macaws pair for life and nest in cavities inside tree trunks, often high up from the ground. The hyacinth macaw also nests in holes in earth banks. Once the eggs are laid, macaws are aggressive toward any potential threat. Even tame captive macaws will defy their owners trying to see what is happening.

A typical macaw clutch contains two to four eggs. The nestling macaw is still naked and blind at a week old. At 4 weeks the wing quills begin

The blue and yellow macaw is one of the few macaws that is still relatively common.

to erupt, the bill darkens and the eyes open. The back then begins to grow feathers, followed by the tail and later the rest of the body and head, the young macaw becoming fully feathered by 10 weeks of age. It does not leave the nest for another 4 weeks, except to sit at the entrance. The parents feed it during this time by regurgitation. At 6 months the young macaw is as large as its parents and resembles them.

With such formidable bills macaws are a match for most small predators. Their main predators are very large birds of prey such as the harpy eagle, *Harpia harpyja*.

Conservation

The macaws' habit of feeding in flocks, combined with their garish colors, made them vulnerable to South American indigenous peoples who traditionally hunted the birds with blowpipes and arrows. However, this persecution had a negligible effect on macaw populations compared with the large-scale cage bird trade of more recent times. Thousands of macaws are illegally caught and exported every year. Even more serious is the forest clearance that has taken place on a huge scale in South and Central America since the mid-20th century.

With the removal of large trees from forests, competition for nest sites increases and crowded conditions occur, dramatically reducing breeding success. Moreover, macaws have an extremely low reproductive rate at the best of times, even in pristine habitat. This means that their populations grow slowly and cannot cope with excessive deforestation and trapping pressures.

Several macaw species became extinct in the Caribbean during the 17th, 18th and 19th centuries. The glaucous macaw, *Anodorhynchus glaucus*, vanished from South America in the early 20th century. Today, eight macaw species are considered to be threatened and nearly all the other species are in decline. The blue-throated macaw (*Ara glaucogularis*) and red-fronted or red-crowned macaw (*A. rubrogenys*) are endangered, while Lear's macaw (*Anodorhynchus leari*) and Spix's or little blue macaw (*Cyanopsitta spixii*) are critically endangered.

World's rarest bird

Spix's macaw is the rarest bird in the world: since the early 1990s there has been only a single bird of this species alive in the wild. It is a male, and lives in open woodland near the Sao Francisco River in northern Bahia, Brazil. About 30 Spix's macaws survive in captivity, and in 1995 a female was released near to the lone male's territory. However, the two birds failed to establish a pair-bond. The future for this species of small blue macaw now looks very bleak.

MACKEREL

MACKEREL ARE DIMINUTIVE relatives of tuna and billfish, such as marlin and swordfish. Like their close relations, mackerel are voracious feeders and their streamlined shape makes them agile swimmers. They are found in temperate and tropical seas worldwide and are highly valued by fisheries.

The common Atlantic mackerel, *Scomber scombrus*, has a plump, tapering body that is blue green on the back and silvery below. The back is patterned with darker ripple marks. There are two dorsal fins, the one in front being spiny, and the pelvic fins are well forward, almost level with the pectoral fins. A line of finlets runs from the second dorsal to the tail fin, with a similar row of finlets on the underside of the body

In the eastern part of its range, the Atlantic mackerel is found from Norway to the Canaries. The same species occurs in the western Atlantic from Chesapeake Bay to the Gulf of Maine. The Spanish mackerel, *Scomberomorus maculatus*, is also native to the western Atlantic and ranges from Cape Cod to Miami and the Gulf of Mexico. Two species of mackerel occur on the eastern Pacific coast of the United States. The Monterey Spanish mackerel, *Scomberomorus concolor*, is confined to the north of the Gulf of California, and is an endangered species. The Pacific Sierra mackerel, *Scomberomorus sierra*, is found in the eastern central Pacific, from southern California to the Galapagos Islands and off the coast of Peru. The chub mackerel, *Scomber japonicus*, is found in the Indian Ocean and the South Pacific, though it is absent from Indonesia and Australia. The Japanese Spanish mackerel, *Scomberomorus niphonius*, inhabits the subtropical and temperate waters of China, the Yellow Sea and the Sea of Japan but also ranges as far north as Vladivostock, Russia.

The pygmy mackerel live in the Indian Ocean and the waters to the west of Australia. They are similar in shape to other mackerel, though they are smaller and deeper bodied. The short mackerel, *Rastrelliger brachyosoma*, grows up to 12 inches (30 cm) long and lives in the Indian Ocean and the western Pacific. The Island mackerel, *R. feughni*, grows to 8 inches (20 cm) and is found in the central part of the Indian Ocean and South Pacific. Finally, the Indian mackerel, *R. kanapurta*, which reaches 13¾ inches (35 cm), ranges from the Indian Ocean and western Pacific through the Red Sea and into the eastern Mediterranean via the Suez Canal. This last species is also known locally as the *kembong*, and is fished widely along the coasts of the Indian Ocean from East Africa to the Malay Archipelago.

The fish commonly known as the horse mackerel is not a mackerel but a member of the family Carangidae. However, its habits are similar to those of the true mackerel.

Marine merry-go-round

Mackerel live in shoals but to a lesser extent than, for example, Atlantic and Pacific herrings (*Clupea harengus* and *C. pallasii*). At the end of October they leave the surface waters, and descend to lie densely packed in the troughs and trenches of the sea bottom. Toward the end of December they spread across the surrounding seabed. At the end of January the mackerel move to the surface, form shoals, and head toward the spawning grounds. One of the main spawning sites is near the edge of the continental shelf.

The period of spawning is from April to August, after which the mackerel move into inshore waters, breaking up into small shoals. They stay there until October, when they go down to the bottom again to repeat the cycle.

Mackerel are swift and maneuverable swimmers. They can retract various fins into grooves on the body, further enhancing their streamlined shape.

The brilliant body coloration of the Atlantic mackerel, a commercially important species, fades after death.

ATLANTIC MACKEREL

CLASS **Osteichthyes**

ORDER **Perciformes**

FAMILY **Scombridae**

GENUS AND SPECIES *Scomber scombrus*

WEIGHT
Up to 7½ lb. (3.4 kg)

LENGTH
Up to 20 in. (50 cm)

DISTINCTIVE FEATURES
Slender body with rounded cross section; second dorsal and anal fins followed by 5 finlets; brilliant blue green back with black vertical bands, curves or spots; white lower flanks and belly with pinkish and gold hues

DIET
Zooplankton, particularly copepods (a type of crustacean); also shrimps, bristle-worms and small fish

BREEDING
Age at first breeding: 2 years; breeding season: April–August; number of eggs: up to 90,000 per spawning; hatching period: 2–6 days; breeding interval: 1 year

LIFE SPAN
Probably up to 20 years

HABITAT
Cold and temperate seas. Winter: relatively deep waters. Spring: moves to shallower waters nearer coast when temperature is 52–57° F (11–14° C).

DISTRIBUTION
Both sides of North Atlantic, including southwestern Baltic Sea, Mediterranean and Black Sea

STATUS
Common

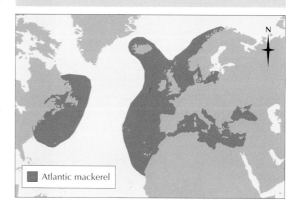

Atlantic mackerel

The female lays up to 90,000 eggs per spawning, each about ⅖ inch (1.2 cm) in diameter. Each egg contains a small oil globule and floats at the surface for 2 days. It then sinks slowly down to mid-water, where it remains suspended for a short while. If the temperature is favorable, about 58° F (15° C), the egg hatches and a larval mackerel about 3.5 millimeters long, still bearing a yolk sac, is born. The yolk lasts for about 9 days, after which the young mackerel begins to hunt minute plankton. Most species of mackerel take 2 years to mature, and grow to about 20 inches (50 cm).

Seasonal changes of diet

On the seabed, mackerel feed on shrimps and smaller crustaceans, marine bristle-worms and small fish. When they return to the surface in January, they change their diet, taking animal plankton, especially the copepod *Calanus*, selectively picking them from the water and snapping them up. From June to October, when they inhabit inshore waters, mackerel take small fish, especially young herrings, sprats and sand eels.

Mackerel hunt mainly by smell. Breton fishers are able to lure them by pouring stale fish blood overboard and scooping up the mackerel attracted to it. At close quarters, however, the fish rely on their sight. With their pelvic fins well forward, they can turn in a tight circle to catch prey that is less maneuverable than themselves.

Important food fish

Mackerel are preyed upon by fast-swimming predatory fish, particularly in their first 2 years of life. To the fishing industry, mackerel are second only to herring and cod in importance among pelagic fish. They are caught in nets, on long-lines, or by spinning, in seine nets from March to June and by hook and line from July to October.

MAGPIE

TWELVE SPECIES OF CROWS are called magpies, but they belong to five different genera and are not closely related. This article concentrates on the two best known species, both of which are in the genus *Pica*. One of these, the black-billed magpie, *P. pica*, has a broad range that encompasses much of Europe and Asia and the western part of North America. The other species, the yellow-billed magpie, *P. nuttalli*, has a restricted range in California.

The name magpie was first used in relation to the black-billed species, which was originally called the pie. The feminine prefix "mag" was added at about the end of the 16th century.

Black bill or yellow bill?

At a distance the black-billed magpie looks entirely black and white but seen close up, especially when the sun is shining, its plumage is shot with iridescent blue and green. It is 17–18

inches (44–46 cm) long, of which more than half is tail. The yellow-billed magpie is slightly smaller but otherwise similar except for its yellow bill and a patch of bare pale yellow skin around each eye.

The black-billed magpie is an eye-catching bird with its long tail and disproportionately short wings. When it lands, the long tail is at once held up and carried clear of the ground. It usually walks, although it often hops sideways with slightly opened wings.

Colorful magpies

A number of the tropical magpies are extremely colorful. The green magpie, *Cissa chinensis*, occurs in southern and Southeast Asia, from the Himalayas south to Sumatra and Borneo. It is bright green with a scarlet bill, a black eye-stripe and reddish chestnut wings. The Sri Lanka or Ceylon magpie, *C. ornata*, is rich blue with brown

Black-billed magpies are highly adaptable birds and seize almost any opportunity to get a meal. In spring they rob other birds' nests of eggs and chicks.

on the head, shoulders and flight feathers. Two species of blue magpies from Southeast Asia, both in the genus *Urocissa*, have very long blue tails with white tips; one has a crimson bill, the other a bright yellow bill.

The azure-winged magpie, *Cyanopica cyanus*, is 13–14 inches (33–36 cm) long with a black cap, sandy brown back, white front and blue wings and tail. Its distribution is most unusual and has long fascinated ornithologists. One population is found in Portugal and Spain, while another is located thousands of miles to the east, in China and Mongolia. Azure-winged magpies may once have occurred in a continuous broad band right across Eurasia, having since died out in much of this range to leave two isolated populations. Another theory is that one of today's populations of azure-winged magpies is the result of artificial introductions by humans.

No food refused

The black-billed magpie lives in many types of open or lightly wooded country, including thickets, scrub, hillsides, orchards, parkland, gardens and well-wooded farmland. It avoids dense woodland, wetlands and rocky areas. Usually the black-billed magpie is seen in one's, two's or three's, but it may form small flocks in winter or when going to roost. It readily takes bright objects and hides them and for this habit is commemorated as a thief in English proverbs.

Although the food of black- and yellow-billed magpies is mainly insects, there seems no limit to what they will take. Their diet ranges

The yellow-billed magpie is restricted to a small range in California, centered on the Sacramento and San Joaquin Valleys. It favors oak groves, streamside trees and cultivated fields.

PICA MAGPIES	
CLASS	**Aves**
ORDER	**Passeriformes**
FAMILY	**Corvidae**
GENUS AND SPECIES	**Black-billed magpie, *Pica pica*; yellow-billed magpie, *P. nuttalli***

WEIGHT
Usually 2¼–2¾ oz. (64–78 g); yellow-billed magpie slightly smaller

LENGTH
Head to tail: 17–18 in. (44–46 cm); wingspan: 21–24 in. (53–60 cm)

DISTINCTIVE FEATURES
Black-billed magpie: strong black bill; black and white plumage with iridescent greenish blue sheen to wings and tail; very long, graduated tail. Yellow-billed magpie: yellow bill; patch of bare yellow skin around eye.

DIET
Invertebrates, fruits, seeds, nuts, bird eggs and nestlings, small vertebrates and carrion

BREEDING
Black-billed magpie. Age at first breeding: 1–2 years; breeding season: eggs laid April; number of eggs: usually 5 to 7; incubation period: 21–22 days; fledging period: 24–30 days; breeding interval: 1 year.

LIFE SPAN
Up to 15 years

HABITAT
Open or lightly wooded country including scrub, thickets, hedgerows, farmland, foothills, riverside trees and urban gardens

DISTRIBUTION
Black-billed magpie: Europe east through much of Asia; western North America. Yellow-billed magpie: central California.

STATUS
Both species common

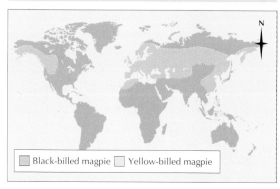

☐ Black-billed magpie ☐ Yellow-billed magpie

from small invertebrates such as snails, slugs and worms, to baby rats, mice and small lizards. They are notorious egg-stealers and also take the nestlings and fledglings of many songbirds. Where game birds are preserved, the black-billed magpie is frequently shot on sight. Nevertheless, elsewhere than in game preserves, it has been found on balance to be beneficial to agriculture because of the insects it takes. Magpies often perch on cattle, sheep and deer to clean up the ticks. Plant matter is also eaten, especially in autumn and winter, and includes such foods as grain, acorns, beechmast, berries, peas and fruits.

Domed nests of thorns

Black-billed magpies begin breeding in April, both partners of a pair combining to build a large nest in a tall tree, thorn bush or overgrown hedge, or sometimes on a telephone pole. The nest is made of sticks cemented with mud and lined with fine roots, dry grass or hair. Typically a domed roof of thorny twigs covers the nest, but this may sometimes be missing. Usually five to seven, rarely up to nine, greenish blue or yellowish eggs are laid. The clutch is incubated by the female alone for 21–22 days. The nestlings

are fed by both parents for about a month on average. Young black-billed magpies have the pied plumage of the parents, but they have short tails. Yellow-billed magpies have similar breeding habits to their black-billed relatives, but often nest in loose colonies.

Ceremonial assemblies

Black-billed magpies are sometimes seen in spring in larger groups than usual, from half a dozen to as many as 100, in what are called ceremonial assemblies. The purpose of these assemblies is not clear but the air of excitement about them is obvious. They probably have something to do with breeding and social status.

Research has shown that these assemblies are initiated by older, high-ranking birds, which deliberately fly into the territory of a neighboring pair to provoke a response. The ensuing displays and squabbles may serve to confirm territory ownership and the status of individual birds. The magpies hop about on the branches of trees or on the ground or take slow flights into the air. They chase each other, raise and lower their head feathers and repeatedly raise their tails, opening and closing them like fans.

Azure-winged magpies are sociable all year and nest in small groups. Some breeding pairs may receive help from younger birds without a mate. These helpers are often last year's offspring.

MAGPIE LARK

Magpie larks (female, above) are noisy, conspicuous birds. The duet sung by mated pairs is one of the best-known sounds of the Australian bush.

THE TWO MAGPIE LARKS belong to a family of their own, Grallinidae. They form part of a larger group of birds from Australia and New Guinea known as mudnest-builders. The magpie lark, mudlark or peewit, *Grallina cyanoleuca*, lives throughout Australia and is the size of a Eurasian blackbird, *Turdus merula*, or American robin, *T. migratorius*. It is neither a magpie nor a lark but was so named because its plumage reminded early settlers of the black-billed magpie, *Pica pica*, from back home in Europe, and because its shape is larklike.

The magpie lark's bold plumage is black and white, being mainly black on the head, neck, back and breast. The female differs from the male in lacking a white eyebrow and in having a white, rather than black, forehead and throat. The species' smaller relative, the New Guinea magpie lark or torrentlark, *G. bruijni*, is confined to the mountains of New Guinea.

Also in the group of mudnest-builders are the apostlebird, *Struthidea cinerea*, and the white-winged chough, *Corcorax melanorhamphos*. The former is larger than the magpie lark and is ashy gray overall with brown flight feathers. The name apostlebird is derived from its habit of associating in small flocks, supposedly of about 12 birds. The white-winged chough is no relation of the true choughs, which are Eurasian crows and belong to the genus *Pyrrhocorax*, although it superficially resembles them. It is 17–18½ inches (43–47 cm) long and is the largest member of the mudnest-builders. It has a slender, downcurved bill and glossy black plumage, with a white patch on each wing.

Communal living

Magpie larks live in open countryside and have adapted themselves to life in the suburbs, where their abundance and flashy plumage have made them popular. They often mate for life and a pair keeps the same territory from year to year. In the autumn and winter some adults join flocks of immature magpie larks numbering several hundreds or occasionally up to 3,000.

Magpie larks in a flock roost together, sometimes on the same site year after year. They also display communally in spectacular feats of flying, which are seen most often when they arrive at and depart from the roost. They fly through the trees in a seemingly mad rush, twisting and turning and sometimes brushing the foliage with their wings. Another habit is for the flock of magpie larks to soar around in circles, weaving in and out, landing, and then repeating the maneuver.

MAGPIE LARK

CLASS	**Aves**
ORDER	**Passeriformes**
FAMILY	**Grallinidae**
GENUS AND SPECIES	***Grallina cyanoleuca***

ALTERNATIVE NAMES
Mudlark; peewit; peewee; pugwall; little magpie; Murray magpie

LENGTH
Head to tail: 10–12 in. (25–30 cm)

DISTINCTIVE FEATURES
Striking black-and-white plumage; pale yellow eye; fairly long, whitish bill; long black legs. Male: black forehead and throat; white panel on side of head below eye. Female: white forehead and throat; broad black band from crown to breast.

DIET
Invertebrates such as grasshoppers, locusts, flies, insect grubs, earthworms, spiders and snails; also small reptiles and frogs

BREEDING
Age at first breeding: usually 2 years; breeding season: after rains in wet season; number of eggs: 3 to 5; incubation period: about 15–18 days; fledging period: 18–21 days; breeding interval: up to 2 broods per year, but dependent on sufficient rainfall

LIFE SPAN
Not known

HABITAT
All types of open country except waterless desert; often in urban areas and beside rivers in trees, swamps and pastures

DISTRIBUTION
Throughout Australia except Tasmania

STATUS
Very common

Magpie lark

Farmer's friends

Magpie larks feed on garden lawns, pastures, swamps and the banks of rivers where they find insects, including locusts and flies, and other invertebrates, such as earthworms. They are beneficial because they eat pests such as cutworms and grasshoppers and the snails that carry flukes that infest cattle. The opening up of the bush in Australia has favored magpie larks because it has provided them with large supplies of insect food. Dung is searched for grubs, and stable flies are caught as they emerge. In northern Australia magpie larks are nicknamed "stock inspectors" because they perch on the backs of domestic animals to remove ticks.

Mud castles

At the start of the breeding season, male magpie larks advertise for mates with *pee-o-wit* calls, which give them one of their popular names. Several males may compete for one female, but once she has been attracted to one male the pair leaves to set up a territory. They sing duets, one bird calling *tee-hee* and the other immediately answering *pee-o-wit*. The territory is fiercely defended against other magpie larks. Passing Australian magpies, *Gymnorhina tibicen*, and birds of prey are also noisily attacked.

Magpie larks begin breeding at the onset of the rains and often nest near lakes or rivers. Much mud is needed for the nest, and in dry years magpie larks may fail to breed. The nest is like an earthenware bowl placed on the branch of a tree. The mud is molded with the breast to form a bowl, and the walls are strengthened by the inclusion of grass, feathers and horsehair. Both parents incubate the three to five eggs. The chicks are fed by both parents and stay in the nest for 2½–3 weeks. Later the young birds form flocks and the parents raise a second brood.

Like the magpie lark, the apostlebird (pictured here) builds mud nests. Both species feed, roost and nest in small groups in which they work together to build bowl-shaped nests and raise their offspring.

MALLARD

A male mallard in full breeding plumage, worn from September to June. Mallard are among the most widespread and numerous of all ducks.

ALTHOUGH THERE ARE many species of wild ducks, the mallard is probably the one that most people in the Northern Hemisphere think of as the "wild duck." It is also the ancestor of most domesticated ducks. The mallard is about 2 feet (60 cm) long and weighs 1⅔–3 pounds (0.75–1.4 kg). The male, or drake, is brightly colored from September to June. His belly and most of his back are gray. His head and neck are a dark glossy green, and a white ring at the base of the neck separates the green from the brown of the breast. He has small, curled feathers on the tail, and his voice is a low, hoarse call. The female, or duck, is a mottled brown, her voice is a loud quack and she has no curly tail feathers.

Eclipse plumage

From July to August the drake mallard is in eclipse plumage. That is, he molts his colorful feathers at the end of June, is clothed in a mottling similar to that of the duck and resumes his colored plumage at the end of August. During this time he is replacing the flight feathers and so cannot fly. At all times of year both sexes have purplish blue specula (wing patches).

Mallard breed in Europe and Asia from the Arctic Circle southward to the Mediterranean, Iran, Tibet and central China, and in northern and central North America. Throughout this huge range there is a general southward movement in autumn to North Africa, southern Asia, the southern United States and Mexico.

Any wetland will do

Mallard are attracted to almost any fresh water, from small ponds in woodlands to large lakes, reservoirs, rivers, irrigation ditches, streams and marshes, although they often live on dry land well away from water. This habit is taken advantage of by wildfowlers and bird-lovers alike as mallard can be encouraged to breed quite easily by digging a pond with small islands or floating basket nests. Mallard spend much time on land even when water is available, standing or sitting about and preening from time to time. On land they waddle apparently awkwardly; on water they swim easily and dive only when alarmed. In the air mallard fly with rapid, shallow wingbeats and with neck outstretched, rising straight off the water in a steep ascent.

MALLARD

CLASS	**Aves**
ORDER	**Anseriformes**
FAMILY	**Anatidae**
GENUS AND SPECIES	***Anas platyrhynchos***

WEIGHT
1⅔–3 lb. (0.75–1.4 kg)

LENGTH
Head to tail: 20–26 in. (51–66 cm);
wingspan: 2⅔–3¼ ft. (0.8–1 m)

DISTINCTIVE FEATURES
Large and heavily built with orange webbed feet and purplish blue specula (wing patches). Breeding male: glossy green head; thin white neck ring; brown breast; gray belly and upperparts. Female and nonbreeding male: mottled brown all over.

DIET
Fresh leaves, shoots, seeds, grain, berries, acorns, insects, worms, tadpoles, frog spawn and small fish; also scraps such as bread

BREEDING
Age at first breeding: 1 year; breeding season: February–October; number of eggs: usually 9 to 12; incubation period: 27–28 days; fledging period: 50–60 days; breeding interval: 1 or 2 broods per year

LIFE SPAN
Up to 25 years

HABITAT
Almost all freshwater wetlands including pools, lakes, reservoirs, rivers, canals, irrigation ditches, marshes and estuaries

DISTRIBUTION
Throughout Northern Hemisphere from Arctic tundra to subtropical zone; some populations move south in winter

STATUS
Very common

Mallard ▢ summer or all year ▢ winter only

Wide choice of food

Mallard feed by day or by night, mainly on leaves, shoots, seeds, grain, berries and acorns, as well as on small animals such as insects and their larvae, worms, tadpoles, frog spawn and small fish. They graze on land like geese and dabble in mud for food, either on land or in the shallows at the edge of water. Mallard also up-end in deeper water to feed from the mud at the bottom.

Mallard belong to the tribe Anatini, or dabbling ducks, and dabbling is one of their main feeding methods. When a duck dabbles its bill in mud, it is using the lamellae (transverse plates) on the inner edges of its bill as a highly efficient filter. As the duck dabbles, its tongue acts as a piston, sucking water or mud into the mouth and driving it out again. Only the edible particles are left behind on the lamellae. Mallard have many taste buds arranged in rows along the sides of the tongue, and these may help it to sort out edible from inedible particles.

Ritual courtship

Mallard form pairs in autumn and begin breeding in early spring. Pairing is preceded by a ritualized courtship. This is initiated by a duck swimming rapidly among a group of drakes with an action that has been called nod-swimming or coquette-swimming. She swims with the neck outstretched and just above water and head nodding. This makes the drakes come together in a tighter group, and they begin their communal displays. These are made up of stereotyped actions known as mock drinking, false preening, shaking, grunt-whistling, head-up-tail-up movements and up-and-down movements.

Mock drinking is a formalized gesture of peace, and two drakes meeting head on will pretend to drink. It is a sign that they have no intention of attacking each other. In false preen-

Broods of 12 or more mallard ducklings are not uncommon, but the mortality rate is high and few survive to adulthood.

ing a drake lifts one wing slightly and reaches behind it with his bill as if to preen. Instead, he rubs the bill over the heel of the wing, making a rattling sound. In shaking a drake draws his head back between his shoulders so the white ring disappears. The feathers on the underside of the body are fluffed out, so the drake appears to ride high on the water. The head feathers are raised so the green sheen disappears and the head rises high so the drake is almost sitting on his tail on the water, and then he shakes his head up and down.

When a drake grunt-whistles, he thrusts his bill almost vertically into the water then throws his head back, scattering a shower of water drops, and as he does this he grunts. The next ritualized action, head-up-tail-up, is self-explanatory. In the up-and-down movement the bill is quickly thrust into the water and jerked up again with the breast held low in the water. Yet another movement is known as gasping; one drake utters a low whistle and the rest give a kind of grunt.

These actions may be made in sequence by a group of drakes facing into the center or by one or two drakes, or between a single drake and duck. Also, any one of these actions may be performed on its own. Together they form a ritual pattern of courtship carried out in the

Mallard ducklings are covered with yellowish down, broken up by large brown patches. They can feed within hours of hatching.

autumn, but actual mating does not take place until spring. In spite of the complicated courtship, however, mallard are highly promiscuous; a drake will mate with a duck while the drake with whom she is paired looks on.

High-diving ducklings

The nest, built by the female, is a shallow saucer of grass, dry leaves and feathers lined with down. It may be on the ground, usually under the cover of bushes or a waterside tree, in the disused nest of a large bird such as a crow, or in a hollow in a tree up to 40 feet (12 m) from the ground. Up to 16, usually 9 to 12, eggs are laid, from February to October. They are incubated by the female alone for 27–28 days. When the ducklings have dried, soon after hatching, the female calls them off the nest and leads them to water or, if the nest is far from water, to a suitable feeding ground. Only rarely is the male still in attendance by this stage and even then he plays no part in the care of the ducklings, which take 50–60 days to fledge.

The enemies of adult mallard include birds of prey and ground predators such as foxes but the main losses are at the duckling stage. Crows, magpies, ermines, raccoons, rats and large fish all attack ducklings. A female may hatch a brood of 12 and in 2 weeks be left with only one.

MALLEE FOWL

THE MALLEE FOWL or mallee hen belongs to a group of fairly large, ground-living birds that do not use their own body heat to hatch their eggs. In this they resemble reptiles, but they use more refined methods than any egg-laying reptiles.

Megapodes: a remarkable family

There are 12 to 19 species of turkeylike or pheasantlike birds in the family Megapodidae, the precise number depending on which scientific authority is followed. They live in the Southeast Asian, Indonesian and Australasian regions and are known variously as megapodes (meaning "big feet"), scrub fowl, brush turkeys, mound-builders and incubator birds. This article concentrates on the mallee fowl, *Leipoa ocellata*, the most completely studied species with the most astonishing nesting habits. Other species include the Australian brush turkey (*Alectura lathami*), the orange-footed or Reinwardt's scrub fowl (*Megapodius reinwardt*) and the tabon or Philippine scrub fowl (*M. cumingii*).

The plumage of megapodes is mainly brown or mottled browns and grays, with some coloring on the neck and breast. These may be slate gray to red, although some species have a bare neck. In the mallee fowl the front of the neck and the breast are decorated with a line of dark feathers.

The tail is usually long and shaped like that of a turkey. The legs, feet and toes are large and strong. Megapodes, like pheasants and related game birds, feed mainly on insects and other invertebrates as well as on seeds and fallen fruits, largely found by scratching the ground.

The distribution of the family is from the Nicobar Islands in the Indian Ocean eastward through the Malay Archipelago to the Philippines and Polynesia, and southward to Australia. The habitat ranges from tropical rain forest to semidry scrub. The mallee fowl itself is found among the dry scrub and eucalyptus woodland of inland Australia, where water is scarce and the bird can go without drinking if necessary.

Automatic incubation

Mallee fowl and other megapodes do not brood their eggs like other birds but incubate them by the heat of the sun, volcanic steam or even huge compost heaps. Some megapodes merely scratch out a pit in the sand, lay their eggs in this and cover them with more sand, leaving them to be incubated by the heat of the surrounding ground. On islands in the Solomons Archipelago and elsewhere, certain megapodes scratch pits where volcanic steam finds its way through the soil, and the heat from this incubates the eggs. Other species simply lay their eggs in cracks in

A male mallee fowl making adjustments to his nest mound to control the internal temperature. The eggs buried inside need to be kept at a constant 91° F (33° C).

The mallee fowl is one of eight threatened species of megapodes. Habitat destruction and the harvesting of eggs by humans are the main threats.

MALLEE FOWL

CLASS	**Aves**
ORDER	**Galliformes**
FAMILY	**Megapodidae**
GENUS AND SPECIES	***Leipoa ocellata***

ALTERNATIVE NAMES
Mallee hen; native pheasant; lowan; gnow

LENGTH
Head to tail: 22–24 in. (56–60 cm)

DISTINCTIVE FEATURES
Turkeylike body with long tail and powerful legs, feet and toes; gray head and neck; whitish gray underparts with black mark down center of breast; subtle pattern of black, gray, white, buff and chestnut bars on upperparts

DIET
Mainly ground-living invertebrates; also herbs, seeds, fallen fruits and some small vertebrates such as lizards

BREEDING
Age at first breeding: 3 years or more; breeding season: highly variable but eggs usually laid September–January; number of eggs: usually 15 to 16; incubation period: 62–64 days when temperature inside nest mound averages 91–93° F (33–34° C); fledging period: chicks independent at hatching; breeding interval: 1–2 years

LIFE SPAN
Not known

HABITAT
Scrub and eucalyptus woodland in arid areas with sandy soils

DISTRIBUTION
Southern and western Australia

STATUS
Vulnerable; increasingly scarce due to land clearance

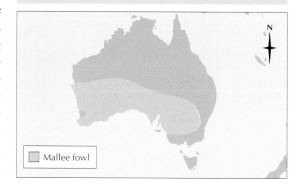

sun-heated rocks. It is in dense rain forests that mound-building reaches its height. There, several pairs of birds build mounds of sand and dead leaves, in varying proportions, up to 35 feet (10 m) in diameter and 15 feet (4.5 m) high. The females dig tunnels into the mound and lay their eggs. The brush turkeys build smaller mounds, 12 feet (3.5 m) in diameter and 3 feet (0.9 m) high, in warm humid forests, where the pile of leaves ferments rapidly and generates much heat. The male tests the mound regularly with his bill, turning the rotting leaves over and mixing them until the correct temperature is reached. Then he allows the female to lay her eggs in the mound before taking charge again.

Incubation in a compost heap

The brush turkeys' mounds work in a way similar to those of the mallee fowl, which has been exhaustively studied. Mallee fowl do not reach maturity until at least 3 years old. Then they pair up and dig a huge pit in the dry sand. There is so little vegetation where they live that they must rake into the pit all the vegetable litter over a radius of 50 yards (45 m). They usually do this in June and July, which is the beginning of the southern winter.

When the winter rains come, the heap of litter is soaked. It begins to heat up with the bacterial action of rotting, and in about August the birds begin to mix sand with the vegetable rubbish in a pit at the center of the mound, to form an incubation chamber. Egg-laying begins in September, by which time the temperature in the incubation chamber should have reached 91° F (33° C). The male tests it with his open bill, the tongue presumably acting as his thermometer, and if the temperature is right, he

allows the female to lay. She also tests the temperature and, if satisfied, scrapes away a small part of the mixture of sand and leaf and lays her first egg in the hole, broad end uppermost. The male replaces the mixture to cover the incubation chamber and then rakes the rest of the vegetable matter over the top to make the mound complete once more.

These actions are repeated at irregular intervals throughout the next 4 months or so. The female lays eggs at intervals of 2 days or more, sometimes as much as 17 days. The male inspects the nest daily to make sure the temperature is correct. The total number of eggs laid may be anything up to 33, each 4 inches (10 cm) long and weighing nearly 8 ounces (225 kg). The female herself weighs only 3½ pounds (1.5 kg) on average, so she is capable of laying more than four times her own weight of eggs in a season. However, a clutch of just 15 to 16 eggs is more usual, which is about twice the average female body weight.

Herculean labors

The temperature of the air varies throughout each day and night and also changes with the seasons. To keep it steady inside the mound means moving and replacing huge amounts of leaves and litter every time there is a rise or fall in the mound's internal temperature. If the mound becomes too hot, the male opens it up at the top to let the heat out. Should the mound show signs of getting too cool, it is opened up still more to let the sun's rays fall directly on the incubation chamber, the material removed being spread beyond the rim of the crater so it can absorb as much heat as possible from the sun. So it goes on, the male regularly testing the temperature and taking the necessary steps to correct it.

No help for the chicks

As soon as the last egg is laid, the first will be hatching. The chicks find their own way to the surface. On arrival there they push their heads out to breathe and rest for a while. Then they heave their bodies out and totter to the nearest shade to rest for a day before starting to look for food. Some die while trying to get out, especially when the mound has been covered with a layer of soil and this has baked hard. The chicks can fly within 24 hours.

Mallee fowl have few natural predators but are plagued by introduced red foxes, *Vulpes vulpes*. One scientist observed more than 70 mallee fowl mounds containing a total of 1,094 eggs. Of these, 15 were broken, 130 failed to hatch and 407 were eaten by foxes. It follows that but for the foxes there would be 90 percent hatching of chicks and this, with the large clutches, suggests a heavy mortality in each generation.

Mallee fowls are often dwarfed by their nest mounds, which reach huge proportions. The chicks struggle free on their own and may never see their parents.

MAMBA

Lightning strikes

The black mamba lives on the ground, sometimes wandering far afield as it hunts for food or seeks a mate. However, it soon returns to a "home," usually a hole in the ground, among rocks or under a fallen tree trunk. The holes are often aardvark burrows or cavities in termite mounds. If threatened or disturbed, the black mamba makes for home, attacking anything that gets in its way.

Besides the relatively high speed with which it moves, the black mamba can strike accurately in any direction, even while traveling fast. It moves with its head raised off the ground, mouth open and tongue flicking. It is also able to expand the neck to form a slight hood and if disturbed gives a hollow-sounding hiss. In striking, it throws its head upward from the ground for about two-fifths the length of its body.

Fast-moving snake

The black mamba is one of the fastest of snakes, with an accurately recorded speed of 7 miles per hour (11 km/h). Speeds of up to about 15 miles per hour (24 km/h) may be possible in short bursts. However, mambas are at a disadvantage on a smooth surface, and black mambas are often run over when crossing roads, especially those with tarred surfaces.

Black mambas will climb into low trees but are more given to climbing rocks, where they lie sunning themselves.

Bird snatcher

The black mamba's prey is almost solely warm-blooded animals, such as birds and small mammals, including rock hyraxes and various rodents. It is said that black mambas are able to snatch birds from the air. They digest food quickly, a large rat being completely digested in 8–10 hours. The green mambas eat birds and their eggs, chameleons, geckos and other tree lizards, as well as small mammals.

Largely tree-living snakes, green mambas (D. angusticeps, above) are less aggressive than the black mamba, with a less deadly venom.

FEARED THROUGHOUT AFRICA for their deadly venom and remarkable agility, the four species of mambas are among the deadliest snakes in the world. All are long, relatively thin, fast-moving snakes. The largest is the so-called black mamba, *Dendroaspis polylepis*, although this snake is actually brown in color. Most large specimens are 8–9 feet (2.4–2.7 m) in length, but bigger individuals are occasionally seen. The largest black mamba ever recorded was 14 feet (4.3 m) long. There are three species of green mambas: *D. angusticeps*, *D. jamesonii* and *D. viridis*. Smaller than the black mamba, they are rarely more than about 6 feet (1.8 m) long.

The black mamba, also called the black-mouth mamba, is the most terrifying of these snakes because its venom is the most toxic. It is found in savanna and bushveld habitats. The green mambas are more often found among dense bushes and in trees.

Mambas are found across most of Africa south of the Sahara. However, the deadly black mamba is confined to the eastern part of the continent, from Somalia and Kenya south to the central part of South Africa.

MAMBAS

CLASS	**Reptilia**
ORDER	**Squamata**
SUBORDER	**Serpentes**
FAMILY	**Elapidae**

GENUS AND SPECIES **Black mamba, *Dendroaspis polylepis*; green mambas: *D. angusticeps*, *D. jamesonii* and *D. viridis***

ALTERNATIVE NAME
Black mamba: blackmouth mamba

LENGTH
**Black mamba: up to 9 ft. (2.7 m).
Green mambas: up to 6 ft. (1.8 m).**

DISTINCTIVE FEATURES
Long, relatively thin, fast-moving snakes; large scales; long front teeth; dark eyes. Black mamba: usually brown in color. Green mambas: bright green.

DIET
Birds and small mammals; also lizards and eggs (green mambas only)

BREEDING
Black mamba. Breeding season: spring and early summer; number of young: 9 to 14; hatching period: 80–90 days; breeding interval: not known.

LIFE SPAN
Up to 14 years in captivity

HABITAT
Black mamba: open savanna and bushland. Green mamba: dense trees and bushes.

DISTRIBUTION
Black mamba: eastern Africa from Somalia and Kenya south to the central part of South Africa. Green mambas: sub-Saharan Africa.

STATUS
Generally common

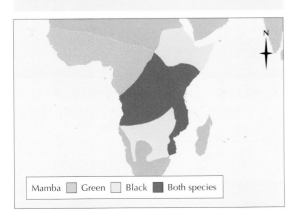

Mamba ☐ Green ☐ Black ☐ Both species

Rapid growth

All mambas lay eggs, those of the green mambas being slightly smaller. The breeding season is usually spring and early summer, but varies from place to place.

A female black mamba lays 9 to 14 oval eggs of about 2¾ inches (7 cm) in length. The eggs are laid in a termite nest or a hollow log. The young snakes hatch in 80–90 days, the young being 18–24 inches (40–60 cm) long at birth. They grow rapidly: one individual was recorded as reaching 6 feet (1.8 m) after only 1 year. A further indication that growth is rapid is that mambas less than 6 feet (1.8 m) long are rarely seen. Young black mambas are grayish green to olive green at birth, gradually getting darker as they grow. Baby green mambas are bluish but become brighter green as they grow.

Striking at soft skin

Mambas' main predators are mongooses, but only while they are young. It has been noticed that when a mamba attacks its prey or is defending itself against, for example, a dog, it strikes deliberately and unhurriedly at places where there is soft skin exposed with a minimum of hair, such as behind the ear, on the cheek below the eye or in the "armpit." A mongoose, with the exception of the face, lacks these vulnerable spots, so has a certain amount of protection against these aggressive snakes. Eagles and secretary birds also kill mambas and young ones may be eaten by snake-eating snakes.

Male black mambas in conflict. The bite of these aggressive snakes is fatal to humans unless antivenin treatment is given.

MANAKIN

Male manakins often display communally to increase their success in attracting females. One male may lead the display and signal its end or start a new sequence.

MANAKINS ARE SMALL, titmouse-sized birds that live in the forests of Central America and South America. They have a stocky build with stout bodies, comparatively large heads and short tails and wings. The males are conspicuous, with splashes of color and sometimes crests or elaborate tails. The females, in contrast, are usually green and unobtrusive.

The red-capped manakin, *Pipra mentalis*, is one of the most striking birds in the forests of Central America, despite its small size. Its plumage is velvety black, and the male has a red head and neck. The bill, eyes and thighs are yellow. The plumage of the male white-bearded manakin, *Manacus manacus*, one of the common forest birds of Trinidad, is black and white with orange legs. The female is olive green and also has orange legs. The golden-headed manakin, *P. erythrocephala*, is probably even more abundant and is certainly more striking. The male is black with a golden orange cap and red and white thighs. The female is olive green.

Rain forest birds

Short wings and short tails are typical of small forest birds, such as manakins. They can be seen flitting through foliage in search of fruits, spiders and insects. The manakins snap up prey while still in flight, then carry it back to a perch to eat. Fruit makes up the bulk of their diet, however, and they have a wide gape, enabling them to swallow fruits ⅔ inches (1.7 cm) across. Manakins also join flocks of other small birds in following army ant hordes to catch the small animals that the ants drive from cover.

Elaborate courtship

Male manakins display for much of the year, leaving their display sites only to eat and drink. Each male has a special display ground, called a court or arena. Gould's manakin, *M. vitellinus*, and the white-bearded manakin clear small patches of ground, about 3 feet by 2 feet (90 cm by 60 cm), of all leaves, twigs and roots. They sometimes display from perches on saplings, as do the red-capped and golden-headed manakins. There may be 50 or more courts in a relatively small area of forest.

Studying the displays, which may involve several males in a "lek" (an assembly ground for display and courtship) is difficult because it is hard to see the rapid movements of the manakins among the foliage. Most displays consist of leaps about the court from perch to perch. The white-bearded manakin jumps from perch to perch, turning around in the air so it is facing the way it came and is ready to jump again. Each jump takes only a fraction of a second. The displays are enhanced by songs and wing noises. The primary flight feathers are narrow and stiff and produce a whirring noise in flight. The shafts of the secondaries are very stout; the vanes are stiff and produce a loud snap during some displays.

Female manakins are attracted to the courting males. As a female arrives, the displays get more intense until she becomes attracted to one male. The males of the blue-backed manakin, *Chiroxiphia pareola*, dance in pairs. When a female arrives, they encircle her. One male flies up and backward while the other lands in its place. Then the second flies up as the first lands again. After mating with one of them, the female flies away. Later she rears the brood by herself.

Females raise young alone

The female manakin makes a small nest in the fork of a branch, at a height of 3–50 feet (1–15 m) above the ground. It is made of dead leaves,

LONG-TAILED MANAKIN

CLASS	**Aves**
ORDER	**Passeriformes**
FAMILY	**Pipridae**
GENUS AND SPECIES	***Chiroxiphia linearis***

LENGTH
**Male: 8¼–9½ in. (21–24 cm);
female: 5⅓–6 in. (13.5–15.5 cm)**

DISTINCTIVE FEATURES
Breeding male: black overall; crimson patch on crown; sky-blue back; orange legs; black bill; central tail feathers are 4–5 in. (10–12.5 cm) long. Female and juvenile: olive overall, paler below; orange legs; dark gray bill; central tail feathers are 1–1½ in. (2.5–4 cm) long.

DIET
Mainly fruits; also insects and spiders

BREEDING
Age at first breeding: probably 3 years (male), up to 3 years (female); breeding season: year-round; number of eggs: 2; incubation period: 17–20 days; fledging period: 13–20 days; breeding interval: not known

LIFE SPAN
Not known

HABITAT
Humid evergreen and semideciduous forest in understory (vegetation growing between forest canopy and ground cover); mainly in lowlands, but also found up to altitude of 4,920 feet (1,500 m)

DISTRIBUTION
Pacific slope of Central America from southeast Mexico south to Costa Rica

STATUS
Common to fairly common

Long-tailed manakin

rootlets and fungi bound together with cobwebs. The female lays one or two eggs and sits on, rather than in, the nest to incubate them. At first she is easily disturbed, but as the incubation proceeds she becomes bolder. Incubation in some species lasts about 17–20 days. The chicks are naked and their eyes do not open for 5 days. Their mother feeds them on insects and fruits. The diet contains a large proportion of insects. This is usual in fruit-eating birds, as the chicks need extra protein for growth. The chicks stay in the nest for 2 weeks, then rest hidden in nearby foliage until they can follow their mother.

The male Gould's manakin often uses saplings as perches during its courtship display. A female may join the display before mating takes place.

Emancipated males

When male birds are brightly colored and the females are drab, the males often have little to do with caring for the young. The males' colors are an advantage in attracting a female but a handicap in keeping the nest hidden from predators. Once the male has given up parental duties, he is freed from a permanent link with the female and develops elaborate plumage and displays. Manakins and other forest birds have small clutches of eggs, which also reduces the labor involved in feeding the offspring.

Males stimulate each other to more dramatic displays, so their communal "attraction" is more effective than that created by the same number of solitary males. However, elaborate displays by solitary males are known in some species.

MANATEE

Manatees must rank among the world's most sedate large mammals. Here a West Indian manatee rests just below the water's surface.

WITH THEIR SPLIT LIPS and hairy, creased faces, manatees make bizarre mermaids, yet some people ascribe the legend to them. There are three species. The West African manatee, *Trichechus senegalensis*, is found along the coast and in certain large rivers of West Africa, and the West Indian manatee, *T. manatus*, lives in the Caribbean from the southeastern United States south to northern South America. Finally, the Amazonian manatee, *T. inunguis*, occurs in the Amazon River.

Like the related dugong, *Dugong dugon*, manatees have large, extremely rounded bodies. They are up to 15 feet (4.5) long from head to tail and can weigh up to 1,320 pounds (600 kg). They

have hairless skin, paddlelike forelimbs and a horizontal tail that is broadly rounded and shovel-like, not notched into two lobes as in the dugong. The flippers are mobile and can be used as hands. Manatees are dark gray to blackish in color. There are further facial differences between the manatees and the dugong: the manatees have no front teeth and the bristly upper lip is divided and mobile, the two opposing halves being used as a highly versatile grasping organ for plucking underwater vegetation.

A life of ease

Singly, or in groups of 6 to 20, manatees swim sluggishly in the sea, in coastal lagoons and in rivers. Inquisitive creatures, they will investigate any strange objects, such as fishing boats, peering at them myopically, their eyesight being poor. When not feeding, they rest at the surface with only the arched back exposed. In shallow waters they may "stand" with the head and shoulders out of water. Adult manatees swim with the tail, using the flippers to turn, but the babies swim using their flippers. Manatees usually surface every 5–10 minutes and take two or three breaths, but they can remain underwater for 16 minutes so long as they are inactive. Manatees communicate with one another by muzzle-to-muzzle contact. They make high-pitched chirps when alarmed.

Feed on aquatic vegetation

Manatees are active at any hour but feed mainly at night on aquatic plants, especially eelgrass, and they will pluck leaves from land plants overhanging the water. The two lip halves can move independently to grip food, the bristles helping push it in, perhaps with some help from the paddles. Manatees seem to eat any vegetation within reach, provided it is not too tough to be pulled apart with the lips. They occasionally eat invertebrates and fish and there are reports that they may take some fish from nets.

Playful courtship

There is no distinct breeding season, but breeding peaks vary between species and geographical location. When the female is in season she is followed by males for 2–4 weeks, although she is receptive only briefly. The female is promiscuous when in heat. A dozen manatees come together and move as a herd into shallow water. There they pair off. The pairs then drag themselves half out of the water and embrace while lying on their sides. After mating they return to the water and play vigorously as a herd.

MANATEES

CLASS	**Mammalia**
ORDER	**Sirenia**
FAMILY	**Trichechidae**
GENUS AND SPECIES	**Amazonian manatee, *Trichechus inunguis*; West Indian manatee, *T. manatus*; West African manatee, *T. senegalensis***

WEIGHT
440–1,320 lb. (200–600 kg)

LENGTH
Head to tail: 6½–15 ft. (2–4.5 m)

DISTINCTIVE FEATURES
Rounded body with small head; square snout with split upper lip (lip halves can move independently); small eyes; no external ear flaps; evenly rounded tail

DIET
Aquatic vegetation; also invertebrates and fish

BREEDING
Age at first breeding: 9–10 years (male); breeding season: all year; number of young: usually 1; gestation period: around 365 days; breeding interval: 2–3 years

LIFE SPAN
Up to 44 years in captivity

HABITAT
Freshwater rivers, lagoons, lakes and backwaters; also shallow coastal waters and swampy estuaries (West Indian and West African manatees only)

DISTRIBUTION
Amazonian manatee: Amazon River Basin, South America. West Indian manatee: southwestern U.S. south to northern South America. West African manatee: West Africa.

STATUS
Amazonian manatee: endangered; other species threatened

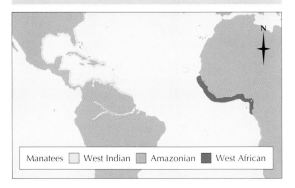

Manatees | West Indian | Amazonian | West African

There is usually only one calf at a birth, occasionally twins. Born underwater, the baby is immediately brought to the surface by the mother to take its first breath. The gestation period is about 1 year. The baby is pink, is 3 feet (90 cm) long and weighs up to 60 pounds (27 kg). It becomes mature at 2–3 years old, when about 8 feet (2.4 m) long, but males may not breed until 9 or 10 years old.

Manatees feed mainly on aquatic vegetation and it was once thought they could be kept to clear ponds and rivers of weed. However, manatees do not tolerate such a life of semi-domestication.

Hunted by humans

Manatees have few natural predators other than alligators. However, they have been killed by local peoples for centuries for their flesh and hides. The Amazonian manatee is now endangered and protected by law in several places in tropical America. Populations of the West African and the West Indian manatees are also declining. Perhaps the biggest menace today is the disturbance of their habitat by the increasing use of outboard motorboats, especially for sport.

Mermaid legend

It is sometimes said that manatees gave rise to the stories of mermaids. The only basis for this is that Christopher Columbus noted in his journal for January 1493 that when off the coast of Haiti he saw three mermaids that rose well out of the water. His opinion of them was that they were not as beautiful as they had been painted, although to some extent they had faces like men. Later Columbus realized they were manatees, animals he had probably encountered before, on the coast of West Africa.

MANDRILL

An adult male and female mandrill and their young. Mandrills live in small family groups comprising a single dominant male, a harem of females and their offspring.

Nobody who has been to a zoo where the mandrill is kept can have failed to have noticed it or to have been impressed by its colorful appearance. The mandrill, *Mandrillus sphinx*, and its close relative the drill, *M. leucophaeus*, are forest-living, baboonlike monkeys. They are sometimes classified in the genus *Papio*, along with the other baboons, but differ strongly in certain respects.

Both the mandrill and the drill are large and powerful with long, doglike muzzles. The mandrill is, in fact, the largest of all monkeys and can weigh as much as 120 pounds (54 kg). However, the male usually averages around 55 pounds (25 kg), while the female is somewhat smaller at 25–26 pounds (11.5 kg). The mandrill's body is thickset, up to 30 inches (75 cm) in the head and body, with a stumpy, 2–3-inch (5–8-cm) tail. Its muzzle is long and deep and has long, thick ridges on either side of the nose. The nostrils are broad and round, and the male has long, sabrelike canine teeth. The male mandrill is particularly brightly colored. He has a dark brown coat with white cheek whiskers, a yellow beard and a crest on the crown of the head. There is a bright scarlet stripe running down the center of the nose, and the ridges on the muzzle are an equally bright purple and blue. His genitalia are similarly colored, and there is also a lilac or purple patch on the rump, on either side of the ischial callosities, or sitting pads, which are pink. The female mandrill is less brightly colored. Her face is grayish black and she does not have the bright colors on the face or rump.

The drill is slightly smaller than the mandrill, being up to 27½ inches (70 cm) long. Its general color is olive brown with a black face, and there are no grooves on the muzzle ridges. The drill has a white fringe around the face.

Mandrills and drills live in central and western Africa. The mandrill occurs in southern Cameroon, Mbini (mainland Equatorial Guinea), Gabon and the Congo. The drill is found from Cross River in Nigeria, south to Cameroon, Equatorial Guinea (including Bioko Island) and Gabon. Both inhabit tropical rain forest; the drill, in particular, is found to quite high altitudes on Mount Cameroon.

Mainly vegetarian diet

Mandrills and drills live in small family groups in the wild, with one male and 5 to 10 females with their offspring making up each group. Although they seem

MANDRILL AND DRILL

CLASS	**Mammalia**
ORDER	**Primates**
FAMILY	**Cercopithecidae**

GENUS AND SPECIES   **Mandrill, *Mandrillus sphinx*; drill, *M. leucophaeus***

WEIGHT
Male: up to 120 lb. (54 kg); average 55 lb. (25 kg); female: average 25 lb. (11.5 kg)

LENGTH
Head and body: 24–30 in. (61–75 cm); tail: 2–3 in. (5–8 cm)

DISTINCTIVE FEATURES
Very powerful monkeys; doglike muzzle, with large canines. Mandrill (both sexes): dark brown, with white cheek whiskers and yellow beard; adult male: purple and blue ridges on muzzle, scarlet stripe down nose and brightly colored lilac or purple rump patches. Drill: olive brown, with black face.

DIET
Fruits, nuts, fungi and invertebrates

BREEDING
Age at first breeding: 5 years; breeding season: all year; number of young: 1; gestation period: 170–180 days; breeding interval: 1–2 years (mandrill); up to 6 years (drill)

LIFE SPAN
Mandrill: up to 46 years in captivity

HABITAT
Dense tropical rain forest from lowlands up to high elevations

DISTRIBUTION
Mandrill: southern Cameroon, mainland Equatorial Guinea, Gabon and the Congo. Drill: Nigeria, Cameroon, Equatorial Guinea (including Bioko Island) and northern Gabon.

STATUS
Mandrill: at low risk. Drill: endangered.

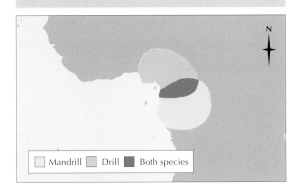

Mandrill | Drill | Both species

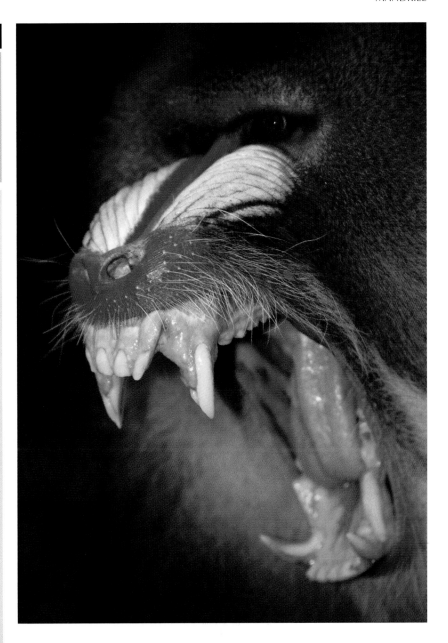

As well as being the largest of all monkeys, mandrills have a quite ferocious appearance: few predators will tackle an adult male.

to spend most of their time on the forest floor, they climb to the middle layer of the trees to feed and to sleep. Like other baboons living on the savanna, they eat mainly plant food together with invertebrates such as worms, termites and ants, and some small vertebrates. Their diet is more vegetarian than that of baboons, however, and fruits, nuts and fungi are important.

Monthly breeding cycle

Mandrills usually bear one baby at a time, although there are occasionally twins. The female is in season every 30 days or so, and this is shown externally by the swelling around her genitalia, which begins after menstruation and reaches a peak at ovulation. Breeding takes place year-round, but there appear to be peak breeding times. The gestation period is 170–180 days and the development of the young is similar to that of the baboons and in monkeys generally. Sexual maturity is reached at about 5 years. In zoos mandrills are long-lived and have been known to live to 46 years. Reproduction is similar in the drill, but births may be spaced as much as 6 years apart. In the mandrill the breeding interval is 1 or 2 years.

Why the bright colors?

Mandrills and drills are not only among the largest of monkeys, but they can also have a quite ferocious appearance. Leopards may take the young from time to time, but they certainly would not attack a full-grown male mandrill. The aggressive appearance is to some extent due to the colors, and zoologists have been hard put to suggest what purpose the coloring serves. Ramona and Desmond Morris, in their book *Men and Apes*, offered one, quite plausible theory. We know that in some monkeys, such as the vervets, the males have brightly colored genitalia, and two males will show these colors to each other as a form of threat display. The Morris's theory is that the mandrill uses its colors in much the same

way, but having a colored face as well provides a double impact of threat toward an opponent. One drawback to this theory is that the drill, so closely related to the mandrill and so like it in almost every other way, has only a black face.

Another theory for the mandrill's color is that it is used as a signal to other mandrills while they are moving through dense bush. The bright colors are easy to spot and make it easier for the monkeys to remain in a tight group.

Hunted for their flesh

Although forest clearing and habitat destruction pose a problem, the main threat to both mandrills and drills is hunting by humans. Both species are hunted for bush meat and it is common for all members of a group, including mothers and their young, to be shot en masse. The encroachment of agriculture has also led to these monkeys being killed in defense of crops. The mandrill is now considered to be at low risk, while the drill is endangered with only 10,000 individuals thought to remain in the wild.

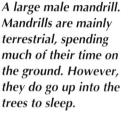

A large male mandrill. Mandrills are mainly terrestrial, spending much of their time on the ground. However, they do go up into the trees to sleep.

MANED WOLF

The maned wolf is a South American fox. It looks rather like a large red fox on stilts because of its very long legs. Its head and body together measure between 3 and just over 4 feet (0.9–1.3 m) and its tail is 1–1⅔ feet (28–50 cm) in length. It stands 2⅖–3 feet (74–90 cm) at the shoulder and weighs up to 57 pounds (26 kg).

Apart from the gray wolf, *Canis lupus*, the maned wolf is the largest member of the dog family. It has a foxlike head and large ears. The coat is shaggy and yellowish red in color, and its legs are dark brown or black for most of their length. There is also some dark coloring around its muzzle, and sometimes on the back and tail. The tip of the tail, chin and throat are sometimes white. The maned wolf gets its name from the mane of long, dark hair on the back of its neck and shoulders to its middle back. The mane is erected in moments of excitement. The maned wolf ranges from the southern parts of Brazil south into Argentina.

Long legs puzzle

The maned wolf is a solitary, mainly nocturnal and extremely swift animal. Like other foxes, it is shy and wary of humans. Not surprisingly, therefore, little is known of its habits. It lives in remote regions, in savanna, grassland and swampy areas, often where small patches of forest are interspersed with open country. It ranges widely over the countryside in search of food.

It was originally thought that the maned wolf had long legs to help it get through the swampy areas that it often inhabits. However, it is now thought that the legs are an adaptation to allow it to look over the long grass found across most of its range. This allows the wolf to see better, and perhaps to hear small rodents in the grass below. It is said that the maned wolf is agile when going uphill but clumsy when descending because its hind legs are slightly longer than its forelegs.

Mice hunters

Maned wolves prey mainly upon small rodents such as mice, which they sometimes dig directly from soft ground. However, at certain times of year they feed mainly on fruits and sugarcane. They also hunt larger rodents such as pacas and agoutis, and take insects, reptiles and birds.

Although not common in zoos, maned wolves have been kept successfully in the Antwerp Zoo in Belgium. One early group was fed on two pigeons and 4½ pounds (2 kg) of bananas a day. Other foods were tried, but lean meat was

Maned wolves have very long, stiltlike legs, which allow them to see better in the long grass found across most of their range.

The maned wolf is vulnerable to hunting because it is believed to pose a threat to livestock. In fact this species eats mainly mice and fruit.

vomited and eggs and milk were refused. Three maned wolves were also kept in the San Diego Zoo and took only bananas at first but took cubed meat later on.

Dark-colored pups

Courtship is similar to that of domestic dogs. Breeding takes place in November in the Northern Hemisphere, or May in the Southern Hemisphere. The pups are born after a gestation period of 62–66 days and there are two to four in a litter. The young have dark brown, nearly black, coats with white-tipped tails. Maned wolves reach sexual maturity at 1 year old and live for up to 15 years in captivity.

Fastest dogs

There are no natural predators on maned wolves. Nor are they hunted for their fur since, having no thick underfur, their pelts are not prized. Nonetheless, these animals are often killed by farmers because it is thought that they are a threat to livestock, attacking lambs, calves and foals. One way maned wolves are hunted is on horseback with a lasso. The species is vulnerable to this because, having no predators, it merely runs for a while and then stops. The idea of continuous flight seems to be absent from its makeup, so it fails to make full use of its speed.

With its long legs, the maned wolf is said to be able to run faster than any other member of the dog family. As a result of hunting there is now a real fear that these animals may soon become extinct in the wild. The maned wolf is presently classified as being at low risk.

MANED WOLF

CLASS	**Mammalia**
ORDER	**Carnivora**
FAMILY	**Canidae**
GENUS AND SPECIES	***Chrysocyon brachyurus***

WEIGHT
44–57 lb. (20–26 kg)

LENGTH
Head and body: 3–4¼ ft. (0.9–1.3 m); shoulder height: 2⅖–3 ft. (70–90 cm); tail: 1–1⅔ ft. (28–50 cm)

DISTINCTIVE FEATURES
Adult: resembles an extremely long-legged red fox; large ears; foxlike head; yellowish red coat; long, dark hair on nape and middle back; dark brown to black lower legs and muzzle; chin, throat and tail-tip sometimes white. Young: dark brown to black coat; white-tipped tail.

DIET
Mainly small rodents, fruits and sugarcane; occasionally medium-sized mammals, birds, reptiles and insects

BREEDING
Age at first breeding: 1 year; breeding season: November (Northern Hemisphere), May (Southern Hemisphere); number of young: 2 to 4; gestation period: 62–66 days; breeding interval: 1 year

LIFE SPAN
Up to 15 years in captivity

HABITAT
Savanna, grassland and swampy areas

DISTRIBUTION
Brazil south to northern Argentina

STATUS
At low risk; estimated population: less than 4,000

Maned wolf

MANGABEY

MANGABEYS ARE SLENDER, long-tailed monkeys, larger and stronger than most monkeys. They live in trees and on the ground and appear similar to typical monkeys, such as guenons and macaques. They have long limbs and rather long, black faces with deep hollows under the cheekbones and white upper eyelids. As in baboons, the skin around the female's genitalia swells up each month as she comes into season. The males have large ischial callosities, or sitting pads. These features, plus their large size, make mangabeys more like the baboons than any other monkeys.

Crested and crestless

There has been, and still is, a great deal of confusion over classification of these monkeys. However, the various species are often separated into two groups, the crestless mangabeys, genus *Cercocebus*, and the crested mangabeys of the genus *Lophocebus*. The first group includes the red-crowned or collared mangabey (*Cercocebus torquatus*), the agile mangabey (*C. agilis*) and the Tana River mangabey (*C. galeritus*). In these species the hair is short and speckled, the lower parts of the limbs are darker than the body and the underparts are white or yellow. These mangabeys have no crest on the head, the eyelids are startlingly white and the tail is slightly longer than head and body combined. There are also a number of subspecies within this group.

The second group includes the gray-cheeked mangabey, *Lophocebus albigena*, and a subspecies of this, the black-crested mangabey, *L. a. aterrimus*. These mangabeys have long, black, unspeckled coats, which are gray on the underparts. They often have ear tufts and always have a tuft on the head. In the black-crested mangabey this stands up in a point and in the gray-cheeked mangabey it falls untidily in all directions. The tail in this group is half as long again as the head and body combined. All mangabeys are 1¼–3 feet (38–90 cm) long, the females being smaller than the males. They weigh between 6½ and 44 pounds (3–20 kg), averaging about 13 pounds (6 kg).

The first group, the mangabeys of the genus *Cercocebus*, live in the rain forests of central-western Africa, from Senegal east to Tanzania and the forest at the mouth of the Tana River,

Kenya. They live on the ground and in the trees. The second group, of the genus *Lophocebus*, is found from western to central Africa, mainly in the tall dense forests of the Democratic Republic of Congo (Zaire), western Kenya and Uganda. These mangabeys tend to be more arboreal (tree-living) than those in the first group.

Distinctive tails

Mangabeys sit and carry their tails in an often characteristic fashion. The gray-cheeked and black-crested mangabeys hold their tails straight up, with the tip usually arched forward above the back when they are standing still or walking slowly. The red-crowned mangabey holds its tail out behind, slightly bowed and curled at the tip when it is running along a branch. When this species of mangabey is standing still, its tail tends to be arched right over the back, with the tip hanging down over the head. When sitting on a branch, the red-crowned mangabey supports

The red-crowned mangabey is one of three species and several subspecies that lack a crest. This group of mangabeys is mainly terrestrial and the animals are often seen on the ground in the forests of western and central Africa.

itself on its hands and feet like most mammals. The gray-cheeked mangabey, however, rests on its rump instead.

Three meals a day

Mangabeys feed mainly on fruits, nuts, seeds and vegetation, but occasionally take small animals as well. Crestless mangabeys have also been known to raid crops.

Mangabeys live in social groups, or troops, of varying numbers of animals. Their day starts soon after dawn. At first the troop merely moves around rather aimlessly, as if still half asleep. They take their first meal 2 or 3 hours after waking and then rest. Another meal is taken about midday, followed by another rest and a third and final meal an hour or two before sunset. Scientifically it is said that mangabeys have three

The black-crested mangabey is distinctive for its black coat, ear tufts and the crest of black hair on its head.

MANGABEYS		
CLASS	**Mammalia**	
ORDER	**Primates**	
FAMILY	**Cercopithecidae**	

GENUS AND SPECIES   **Tana River mangabey,** *Cercocebus galeritus*; **agile mangabey,** *C. agilis*; **red-crowned mangabey,** *C. torquatus*; **gray-cheeked mangabey,** *Lophocebus albigena*; **black-crested mangabey,** *L. a. aterrimus*

ALTERNATIVE NAME
Red-crowned mangabey: collared mangabey

WEIGHT
6½–44 lb. (3–20 kg)

LENGTH
Head and body: 1¼–3 ft. (38–90 cm); tail: 1⅖–2½ ft. (43–75 cm)

DISTINCTIVE FEATURES
Large, slender body; long limbs and tail; usually grayish brown, dark gray or black, with paler underparts; prominent crest of black fur (adult *Lophocebus* only).

DIET
Mainly fruits, nuts, seeds and vegetation; occasionally small animals

BREEDING
Age at first breeding: 4–5 years; breeding season: all year; number of young: 1; gestation period: 165–175 days; breeding interval: not known

LIFE SPAN
Up to 20 years or more

HABITAT
Tropical forest and mangroves

DISTRIBUTION
Tana River mangabey: Kenya and Tanzania. Other *Cercocebus* species: West Africa. *Lophocebus*: western and central Africa.

STATUS
Endangered, threatened or near-threatened

Mangabeys

feeding peaks during the day. Guenons, which often feed with the mangabeys, have only two peaks. This prevents the two kinds of monkeys from getting in each other's way at feeding times.

The female mangabeys groom the males a great deal, especially during the afternoon when the troop is resting. As with baboons, the males rarely or never groom the females.

Hairless babies

Mangabeys breed at all times of the year. The female has a monthly sexual cycle, as in the baboons. They give birth to a single young after a gestation period of 165–175 days. Almost hairless at birth, the baby mangabey clings to its mother's belly fur for transportation. Some of the adult males show a fair amount of interest in the infants, sometimes carrying them around and cuddling them. Once independent of their parents, the young tend to associate in groups and avoid the adults.

Varied social behavior

It is not possible to say more about the social behavior of mangabeys with any degree of confidence because apparently it can vary from species to species as well as from place to place. For example, in Mbini (formerly Rio Muni), in Equatorial Guinea, gray-cheeked mangabeys live in groups of between 9 and 11 individuals with only one adult male in each group. The same species in Uganda lives in groups of 15 to 20, with four or five adult males in each group. This Ugandan group dynamic is much the same as in the red-crowned mangabey. Where there is more than one male in a group, they tend to avoid each other and may wander off on their own for periods of time.

Several ways of meeting danger

Probably the main predator of forest monkeys in Africa is the crowned eagle, *Stephanoaetus coronatus*. This is certainly true for mangabeys. Leopards and pythons take their toll as well.

The behavior of the different mangabey species in moments of danger varies a good deal. Red-crowned mangabeys will stop calling and stay very quiet, but gray-cheeked mangabeys call all the more when threatened or disturbed. If the threat continues, the gray-cheeked mangabey will rapidly escape into the tallest trees. The red-crowned mangabey, on the other hand, runs across the top of the tree canopy, or goes down to the ground, in both cases moving into swamp forest for protection. Mangabeys use their exceptional leaping ability to escape danger, being able to jump as much as 20 feet (6 m), when moving around on branches.

However, the main threat to the mangabeys is loss of habitat due to commercial logging, human encroachment and disturbance. The Tana River mangabey is now endangered as a result of this, while other forms, such as the red-crowned mangabey, agile mangabey and black-crested mangabey, are threatened or near-threatened.

The Tana River mangabey, found at the mouth of the Tana River in Kenya, and in parts of Tanzania, is now an endangered species, mainly as a result of habitat loss.

MANNIKIN

The red-headed finch lives in the savanna and thornscrub of southern Africa. Like other mannikins, it is a seed eater.

MANNIKINS ARE A GROUP of small seed-eating finches in the large subfamily Estrildinae. They are 3¾–4½ inches (9.5–11.5 cm) long, according to species. The largest is the Java sparrow, *Padda oryzivora*, discussed elsewhere. There are about 40 species of mannikins, found mainly in Southeast Asia and Australasia, with some species in Africa, the Middle East and the Indian subcontinent. One species, the Madagascar mannikin or bib-finch, *Lonchura nana*, is found on Madagascar.

Somber plumage

Mannikins are generally somberly colored, often with patches of brown, black and white. However, the cut-throat finch, *Amandina fasciata*, and red-headed finch, *A. erythrocephala*, have red on their heads. Two very common species, which are popular as cage birds, are the spotted mannikin, *L. punctulata*, and white-rumped munia, *L. striata*. The former, also known as the spotted munia, nutmeg finch and spice finch, has a chestnut brown back with white streaks. Its head is reddish brown, the tail grayish yellow and the underparts white with faint brown markings. Several color variations of the white-rumped or striated munia have been produced in

captivity by aviculturists, including one form known as the Bengalese finch. The chestnut-breasted mannikin, *L. casta-neothorax*, has a black head, rich brown back, bright chestnut breast and white belly. Other mannikins include the magpie mannikin (*L. fringilloides*), three species known as silverbills (all in the genus *Lonchura*) and the bronze mannikin (*L. cucullata*).

Sociable birds

The typical habitat of mannikins is open country, especially near open water. The chestnut-breasted mannikin of Australia and New Guinea, which lives in reeds and grasses alongside marshes and streams, has been encouraged in its spread by increased artificial irrigation. It also frequents fields of rice and sugarcane.

Mannikins are small, highly sociable birds, living in flocks and eating, roosting and preening together. The spotted mannikin has three calls, each used for a different purpose by members of a flock. As they fly about they utter a quiet *chip* to keep in touch with each other. A louder *kit-tee* is used for identification and is heard as the birds go to roost. There is also a sharp creaking alarm call. Flocks of mannikins may number several hundred birds, especially outside the breeding season, and often contain several different species. They roost in reeds or stands of tall grass.

Seed eaters

Mannikins live in open country because they are basically seed eaters, feeding mainly on the seeds of grasses. They have several different methods of feeding. The spotted mannikin, for example, hops about on the ground, feeding on fallen seeds. It climbs tall grasses, drawing them together to get at the seed heads, and it hangs from vertical twigs to eat seeds on grass heads below. Along with other mannikins, the spotted mannikin is often a pest in fields of rice and other crops. The chestnut-breasted mannikin also catches termites on the wing, especially in the breeding season, to provide protein for its chicks.

Courtship displays

The first stage in courtship is for the two mannikins to pick up a blade of grass apiece and fly or hop about excitedly with the grass in their

CHESTNUT-BREASTED MANNIKIN

CLASS	**Aves**
ORDER	**Passeriformes**
FAMILY	**Ploceidae**
SUBFAMILY	**Estrildinae**
GENUS AND SPECIES	***Lonchura castaneothorax***

ALTERNATIVE NAMES
Chestnut-breasted munia; chestnut-breasted finch; chestnut finch; barleyfinch; barley sparrow; barley bird

LENGTH
Head to tail: 3¾–4 in. (9.5–10 cm); wingspan: 5–5½ in. (13–14.5 cm)

DISTINCTIVE FEATURES
Short, stout, conical bill; black head; rufous brown upperparts; chestnut breast band; white belly with thick black bars on flanks

DIET
Grass seeds; also insects such as flying termites in breeding season

BREEDING
Age at first breeding: 1 year; breeding season: December–May (north of range); number of eggs: usually 5 or 6; incubation period: 15–18 days; fledging period: about 21 days; breeding interval: 1 year

LIFE SPAN
Probably up to 8 years

HABITAT
Grassland and scrub alongside marshes, mangrove swamps and streams; also roadsides and fields of sugarcane and rice

DISTRIBUTION
Coastal regions of northern and eastern Australia; New Guinea; introduced to some South Pacific islands

STATUS
Common

Chestnut-breasted mannikin

bills. The courtship dance of the spotted mannikin is performed with the male alighting by the female and bowing to her after they have dropped their pieces of grass. He sings a feeble twittering jingle, with tail bent down and feathers fluffed. He also sways from side to side and bobs up and down. The chestnut-breasted mannikin has a similar courtship but both birds dance, repeatedly bowing to each other and then stretching upward.

Large flocks of chestnut-breasted mannikins roam the coastal grasslands of northern and eastern Australia, feeding and roosting together.

Grass dome for a nest

The nest is built in trees or sometimes on buildings, and several may be placed close together to form a colony. It is globular or domelike, about 8 inches (20 cm) long and 6 inches (15 cm) in section, with a chamber inside connected to a short entrance tunnel covered by a porch. The materials used are grasses and bark, with a lining of soft grasses, and because the materials are laid in orderly layers, the walls are strong and waterproof. The spotted mannikins also build special roosting nests in which a crowd of them gather for the night; this provides additional protection from predators.

The clutch varies from three to seven eggs, according to species and environmental conditions, and the eggs are incubated by both parents for about 15–18 days. The chicks grow rapidly and are fully independent in a few months.

Spreading to new regions

Mannikins often escape from captivity and in this way many species have extended their ranges. The spotted mannikin, for example, is native to Asia, from India east to Taiwan and south to Indonesia. Escapes from aviaries have colonized Australia, and it has also been introduced to Hawaii, Mauritius and the Seychelles.

MANTIS

THE NAME MANTIS IS derived from a Greek word meaning "prophet" or "soothsayer" and refers (as also does the epithet "praying") to the habitual attitude of the insect. When at rest it stands motionless on its four hind legs with the forelegs raised and folded in front of the face as if in prayer. Also distinctive are the strongly spined, raptorial forelegs. The joint called the tibia can be snapped back against the femur, rather as the blade of a penknife snaps into its handle, to form a pair of grasping organs that seize and hold the insect's victims. The mantis also has a narrow prothorax (the front part of the thorax), which forms a movable neck.

Mantises, or praying mantises as they are often called, are carnivorous and feed mainly on other insects. They are found mostly in tropical or subtropical countries, although some species occur in temperate regions. Most species measure about 2 inches (5 cm) in length, although they may be smaller or larger than this. They have narrow, leathery forewings and large fan-shaped hind wings, which are folded beneath the forewings when not in use. Most mantises can

fly, but they do not readily take to flight and seldom go far. About 2,000 species are known, one of the most familiar being the European mantis, *Mantis religiosa*. It is found in the Mediterranean region and has been introduced into eastern North America. One of the most widely distributed species in North America is *Stagmomantis carolina*. The largest is another introduced species, the Chinese mantis, *Tenodera aridifolia sinensis*. Native to many parts of Asia, it is 2¾–4 inches (7–10 cm) in length.

Hidden terror

Most mantises spend their time sitting still among foliage or on the bark of trees, waiting for insects to stray within reach of a lightning-quick snatch of their spined forelegs. Nearly all are shaped and colored to blend with their surroundings. Many are green or brown, matching the living or dead leaves among which they sit, but some have more elaborate camouflage, which serves two purposes. First, because they do not pursue their prey but wait for it to stray within reach, they need to stay hidden. Second, their

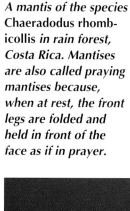

A mantis of the species Chaeradodus rhombicollis *in rain forest, Costa Rica. Mantises are also called praying mantises because, when at rest, the front legs are folded and held in front of the face as if in prayer.*

MANTISES

PHYLUM	**Arthropoda**
CLASS	**Insecta**
ORDER	**Dictyoptera**
SUBORDER	**Mantodea**
FAMILY	**Mantidae**

GENUS AND SPECIES **About 2,000 species, including orchid mantis, *Hymenopus coronatus*; European mantis, *Mantis religiosa*; and African devil's flower, *Idolum diabolicum***

ALTERNATIVE NAMES
Mantid; praying mantis; praying mantid

LENGTH
Average up to 2 in. (5 cm)

DISTINCTIVE FEATURES
Strongly spined, grasping forelegs; narrow prothorax (front part of thorax) forms moveable neck; at rest, legs held folded in front of face; most species green or brown and well camouflaged; some tropical species brightly colored with unusual outgrowths

DIET
Almost entirely insects such as flies, butterflies and grasshoppers

BREEDING
Female often eats male after mating. Number of eggs: 100 to 200; young passes through up to 12 stages before becoming adult.

LIFE SPAN
Varies widely

HABITAT
Forest, woodland and other vegetated areas

DISTRIBUTION
Mainly in Tropics and subtropics; some species in U.S. and other temperate regions

STATUS
Common

grasping forelegs, although formidable to other insects, are, except in the very large species, useless against birds and lizards. Since mantises are slow-moving, they must remain concealed to avoid being caught and eaten.

Mantises never take plant food. They seize insects such as flies, butterflies and grasshoppers in the viselike grip of their spined forelegs and eat them alive, neatly and delicately. Some of the largest species occasionally catch and eat small birds and lizards in the same way.

Females eat males

To the female of most mantis species, the male is no more than another piece of food. He must therefore be careful in his approach if he wishes to mate, rather than be the next meal. On seeing a ripe female, the male freezes, then starts to creep up on her with movements almost too slow for the eye to follow, sometimes taking an hour or more to move 1 foot (30 cm). Once within range, he makes a short hop and clasps the female to mate. If the pair is disturbed, or the female sees her suitor, she will eat him, starting by biting off his head. Because mating in mantises is a reflex, the male continues to copulate even while being eaten. In most cases the male will survive mating, but is often eaten by the female afterward. This is not the practice of all species, however. Some of the smaller mantises, such as *Ameles spallanzania*, are not cannibalistic.

Most mantises, such as this European mantis, blend perfectly into their surroundings by virtue of their shape and coloration.

Cocoonlike capsules

The female mantis lays 100 to 200 eggs at a time in egg sacs called oothecae. Oothecae are formed in various ways, depending on species. They may, for example, be made of a secretion that the female exudes and stirs into a froth by movements of her body. The eggs become enclosed in this while it is still plastic, then it quickly hardens and dries into a tough, spongy material. The oothecae are attached to twigs and branches or hidden under stones. Female mantises produce a dozen or more of these during their lifetime.

The young mantises hatch together and are wormlike in appearance. At first they hang from the egg capsule by silken threads that they extrude from the hind end of the abdomen. However, they soon shed their skin and emerge as tiny mantises, or nymphs. The nymphs grow by gradual stages, or instars, molting up to 12 times before becoming adults. The wings, tiny at first, grow with each succeeding molt.

The egg capsules are a protection against insectivorous mammals, birds and adverse weather conditions, but are no protection against parasitic wasps of the ichneumon type. These are probably the most serious predators on mantises.

Fatal flowers

Some tropical species are even more deceptive in their camouflage, being brightly colored with bizarre outgrowths from their limbs and body. In this way they take on the appearance of flowers and so lure insects such as bees and butterflies within reach. For example, the orchid mantis, *Hymenopus coronatus*, of Malaysia and Indonesia, is colored pink in its young or subadult stage, and the thigh joints of the four hind legs are widely expanded so they look like petals, while the pink body resembles the center of the flower. When the mantis reaches the adult stage, however, its body becomes white and elongated and its resemblance to a flower is largely lost. The African devil's flower, *Idolum diabolicum*, has expansions on the thorax and the forelegs that are white and red. It hangs down from a leaf or twig and catches any flies or butterflies attracted to it.

When they are frightened, many mantises suddenly adopt a menacing posture, rearing up and throwing their forelegs wide apart. One African species, *Pseudocreobotra wahlbergi*, spreads its wings, on which there are a pair of eyelike markings, so the predator is suddenly confronted with a menacing "face."

A praying mantis in Irian Jaya, Indonesia. Mantises are voracious carnivores, feeding on a wide range of other insects. It has been said that preying mantis might be a better name.

MANTIS FLY

LIKE SO MANY INSECTS loosely called flies, the mantis flies are not related to the true, two-winged flies of the order Diptera. Rather they belong to the order Neuroptera, to which the alder flies and lacewings also belong. Mantis flies are so named because of the superficial resemblance of the adults to the mantises, order Dictyoptera. However, mantis flies are not related to the mantises.

Mantis flies are fairly small insects, rarely more than 1 inch (2.5 cm) long. They live mainly in the Tropics and subtropics. The raptorial front legs, in particular, are similar to those of the mantises, being adapted for seizing small insects. In both the mantis flies and the mantises the front limbs work rather like a penknife blade, with the end joint snapping back onto the one behind. Unlike the mantises, however, the wings of the mantis flies are delicate and transparent, with a fine network of veins. This is a characteristic of the alder flies and the lacewings, too. In addition, a mantis fly's forewing is normal, not a leathery covering for the hind wing, as in a mantis, and the hind wing is not folded and pleated. Mantis flies are not only distinctive in appearance, but have a strange and highly specialized life history.

Food and lodging provided

The larvae of the mantis flies are parasites on wolf spiders and wasps, or other insects, depending on species. Adult mantis flies live among foliage, preying on small insects.

Like green lacewings, the female mantis fly lays her eggs at the ends of threads or stalks of silk that harden on contact with the air. A single brood is laid each year.

The southern European species, *Mantispa styriaca*, is one of the few species that has been studied in any great detail. Its eggs are rose red in color and look rather like tiny pins in a pincushion once they have been laid. The larvae that hatch from them are said to be "campodeiform," which means "like a *Campodea*," one of the genera of bristletails. In fact, the larvae resemble small silverfish or tiny earwigs, as do

Adult mantis flies prey upon a variety of small insects, seizing their victims in specialized front limbs. This photo shows a mantis fly of the genus Mantispa, *having just captured its helpless prey.*

MANTIS FLIES

PHYLUM	**Arthropoda**
CLASS	**Insecta**
ORDER	**Neuroptera**
SUBORDER	**Planipennia**
FAMILY	**Mantispidae**

GENUS AND SPECIES **European mantis fly,** *Mantispa styriaca*; **Brazilian mantis fly,** *Symphasis varia*; **many others**

ALTERNATIVE NAMES
Mantispid; mantid fly

LENGTH
Most species: up to 1 inch (2.5 cm). European mantis fly: ⅖–⅘ inch (1–2 cm).

DISTINCTIVE FEATURES
Adult: grasping front legs, similar to those of mantises; delicate, transparent wings with fine network of veins. Larva: 2 forms, first has well-developed legs and squarish head; second is fat, white grub with short legs.

DIET
Adult: small insects. Larva: parasitic on wolf spiders, wasps and other insects.

BREEDING
Usually 3 larval instars (stages); parasitic larva seeks out egg cocoon of host and pupates within it; 1 brood per year

LIFE SPAN
Usually 1 year

HABITAT
Wherever there are suitable prey and hosts

DISTRIBUTION
Mainly in Tropics and subtropics

STATUS
Common

Mantis flies have an unusual life history in that the larval forms are parasitic on the eggs and young of other insects. There are also two larval forms before pupation and the eventual emergence of the adult.

the larvae of other Neuroptera. They go into hibernation immediately after hatching, becoming active again the following spring.

As they become active again, the European mantis fly larvae start to look for the egg cocoons of wolf spiders, genus *Lycosa*. The female wolf spider spins a white, silken cocoon for her eggs and carries it with her wherever she goes, fastened to the tip of her abdomen. A single *Mantispa* larva enters each cocoon, feeding on the young spiders as soon as they hatch. The mantis fly larva becomes very fat and swollen as it devours the entire brood of spiders. The mother spider notices nothing as this happens, continuing to carry the egg sac about with her.

Multiple metamorphosis
The mantis fly larva then molts its skin and reappears in a caterpillar-like or "cruciform" shape, with a relatively tiny head and small legs, looking wholly unlike the original hatchling. Still within the spider's egg sac, it spins a cocoon among the dried remains of its victims and pupates. Just as the larva exists in two distinct and dissimilar stages, so does the pupa, which, as well as an inactive stage inside the cocoon, has a second, active stage outside it.

Insects in which the larva or pupa shows more than one form in the course of development are said to undergo hypermetamorphosis. This is a metamorphosis beyond the normal number of stages. The Brazilian mantis fly, *Symphasis varia*, has a similar life history, but it is parasitic on wasps rather than wolf spiders and its larvae pupate in wasp nests.

Variations on a theme
The development of the raptorial front legs has taken place independently in the mantis flies and the mantises. It is also found in the New Zealand wingless fly, *Apterodromia evansi*. It has forelegs in which two of the joints are developed almost exactly as in a mantis. Mantis shrimps also have spined, grasping claws of this kind. Taken together, these make up an impressive case of convergent evolution, in which a similar feature is developed independently for a particular purpose by several unrelated animals.

MANTIS SHRIMP

THE MOST STRIKING FEATURE of these lobsterlike shrimps is the pair of strong grasping claws, somewhat like those of a mantis. It is from these that the mantis shrimps derive their name. The last joint of each limb closes like the blade of a jackknife on the preceding joint. In some species this last joint is armed with sharp spines, while in others the blade has a smooth cutting edge fitting into a groove in the preceding joint.

There are about 350 known species of mantis shrimps, ranging from 1½ inches (4 cm) to over 1 foot (30 cm) in length. One, *Squilla mantis*, reaches about 8 inches (20 cm) and is found in the Mediterranean and nearby regions of the Atlantic. The most common species on the Atlantic coast of North America is *S. empusa*.

Numerous limbs
The jackknife claws are second in a series of limbs of various kinds on the thorax. They are preceded by a small, slender pair of limbs and are followed by three much smaller pairs of clawed limbs used in digging. Three unclawed, slender pairs of limbs follow these, and on the abdomen are five pairs of broad, flat swimmerets (unspecialized appendages) with filamentous gills. The swimmerets have small hooks that link them in the midline of the body. The final pair of limbs, as in a lobster, forms the sides of a tail fan.

In some species, the abdomen is flattened and widens toward the hind end. Moreover, the abdomen looks longer than it really is because the dorsal shield, or carapace, does not cover the last four segments of the thorax as it does in a lobster. The front of the head has two hinged parts, one carrying the eyes, which are on stalks, and the other carrying the antennae.

Bright colored, marine shrimps
Mantis shrimps are usually brightly colored, often green or brown. Others are blue or red in color. Some may be mottled while others have alternating bands of color.

Mantis shrimps are also known as split-thumbs, squillas, prawn killers, shrimp mammies and nurse shrimps. All are marine and are particularly common in warm, tropical seas. They are usually found in shallow waters at depths of less than 330 feet (100 m), but may be found as deep as 4,300 feet (1,300 m).

Marine guillotines
Mantis shrimps seize their food with their claws, holding it fast with the spines when these are present. They may also stab at prey with these sharp appendages. Mantis shrimps are unusual for crustaceans in being truly carnivorous, that is eating only live prey. Many other supposedly carnivorous crustaceans are really scavengers. Many species rarely leave their burrows but lie in wait at the entrance for their prey. Mantis shrimps mainly feed on small fish, shrimps, crabs, worms, mollusks and anemones. Other species venture out more often in pursuit of food, propelling themselves with their swimmerets.

Mantis shrimps live mainly in shallow water or on the shore, making deep vertical or sloping burrows in the sand or mud, or concealing themselves in holes and crevices in rock or coral. One species, *Gonodactylus guerinii*, which lives in coral, plugs the entrance to its hole with its specially modified and spiny tail fan. The spines not only give protection, but also make the visible portions of this species look like a sea urchin attached to the surface of the coral.

Devoted mothers
In many species the male and female share a burrow or den for up to a week in order to mate. The male is finally evicted after mating. When

Mantis shrimps are often brightly colored, such as this one of the species Odontodactylus scyllarus, *shown under a flashlight in Papua New Guinea.*

A mantis shrimp of the genus Lysiosquilla *off the coast of Papua New Guinea. Unlike most crustaceans, which are scavengers, mantis shrimps are truly carnivorous.*

MANTIS SHRIMPS

PHYLUM **Arthropoda**

CLASS **Crustacea**

SUBCLASS **Malacostraca**

ORDER **Stomatopoda**

GENUS AND SPECIES **About 250 species, including *Squilla desmaresti*, *S. mantis* and *S. empusa***

ALTERNATIVE NAMES
Split-thumb; squilla; prawn killer; shrimp mammy; nurse shrimp

LENGTH
Usually 1½–13¾ in. (4–35 cm)

DISTINCTIVE FEATURES
Slightly flattened body; somewhat resembles a lobster or shrimp, but with reduced carapace (dorsal shield) leaving more of thorax exposed; jackknife-like claws on second pair of limbs, often with sharp spines

DIET
Live prey such as small fish, shrimps, crabs, worms, mollusks and anemones

BREEDING
Varies with species. Eggs may be brooded beneath the thorax or guarded in the den.

LIFE SPAN
Up to 30 years

HABITAT
Usually warm seas in shallows, generally less than 330 ft. (100 m). Lives in burrows in mud and sand; also in holes and crevices in rock and coral.

DISTRIBUTION
Virtually worldwide in temperate and tropical waters; most species in Tropics

STATUS
Varies according to species

mating, the male places his sperm in special pockets on the underside of the female's thorax, near the pair of openings on the sixth segment where the eggs will be laid. The eggs are small, and are glued together by a cement produced by glands near these openings. They stick together in a mass that the female carries on the three pairs of small, clawed limbs and constantly turns over and cleans. In many species, the female mantis shrimp lays the eggs in her burrow, where she guards them until they hatch.

After hatching, the larvae swim or drift about, molting at intervals before eventually becoming bottom-dwelling adults. Some of the larvae in tropical seas are over 2 inches (5 cm) long with a striking transparent, glassy appearance. In some places they are so numerous that the sea looks like a thick soup.

Gladiatorial fights

When fighting, mantis shrimps lash out with their tails and legs but seldom do each other serious injury. A fight is always preceded by a threat display in which the two shrimps spread their legs and claws, exposing certain white spots and silver streaks. It is something of a puzzle why some mantis shrimp should be so heavily armored, yet live in burrows they rarely leave. It seems, however, that the armor is not so much a defense against predators as a means of defending a territory. Sometimes the lawful owner of a burrow is ejected, but only after a lengthy fight.

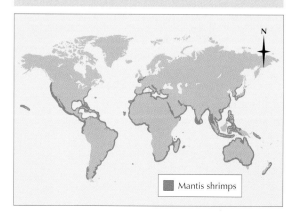

Mantis shrimps

MARABOU

With its massive, heavy bill and pink, almost naked head and neck, the marabou is one of the less attractive storks. It is also the largest stork and can reach 5 feet (1.5 m) from head to tail with a wingspan of up to 9½ feet (2.9 m). The male is, on average, larger than the female. The back, wings and tail are dark gray in color and the underparts are off-white. The legs are also mainly gray. The marabou has a white ruff around its neck and a pink, fleshy, inflatable pouch that hangs down 10–14 inches (25–35 cm) in front of the neck. This is apparently part of the respiratory system, but has now become a secondary sexual characteristic.

The marabou is found in Africa from Sudan and Ethiopia to Zambia, being especially common in East Africa. Closely related to it are the adjutant stork, *Leptoptilos dubius*, and the lesser adjutant stork, *L. javanicus*, of Asia, which range from India to Borneo. The name adjutant is also sometimes given to the marabou, possibly because of the stork's pompous, strutting gait.

Scavenging carrion and garbage

Marabous are in many ways similar to vultures, both in appearance and behavior. The naked head and neck appear to be the same adaptation for carrion-eating as in vultures and condors, as feathers would become matted with blood while feeding on a large carcass. Marabous and vultures often mix around carcasses of big game, but marabous also haunt human settlements where there are carcasses of domestic stock or smaller refuse. They also flock around slaughter-houses and rubbish dumps. On the shores of the large East African lakes they gather near fishing villages to feed on offal left over from cleaning the catch. Because they scavenge garbage and waste, marabous are allowed to roam in cities and in some places they are protected by law.

Scavenging brings marabous into contact with other carrion feeders such as vultures and hyenas. It is quite a common sight to see all three together at a carcass. The marabous dominate the vultures, stealing choice pieces of flesh. Large pieces of bone may also be swallowed. In terms of live prey the marabou is the chief predator of flamingos. It also preys on large insects, fish, frogs, rats and reptiles such as snakes and lizards.

When seeking carcasses of wild game, marabous may have to fly long distances. Like many other large, heavy birds, they do this by soaring in thermals (rising currents of warm air). This is also the method used by vultures and other raptors (birds of prey) that need to conserve

energy. They can glide effortlessly over long distances, but this method restricts takeoff until the sun is quite high and the air has warmed up.

Breeding when animals die

Unlike most tropical birds, which breed in the wet season when there are abundant plants or insects on which to feed their young, marabous breed in the dry season. The advantage of this is that during the dry season their food is concentrated in a few places. For example, the lakes and rivers are low, so fish and frogs are easily caught. Game also concentrates around waterholes, where it may be killed by predators. In agricultural areas, more cattle are killed in the dry season because of the shortage of food and because plowing makes it easy for the marabous to catch mice. Marabous nest within easy soaring and gliding distance of such places, ensuring an easy food supply for the chicks.

An adult marabou with two young. There might be several years between broods, with only about 20 percent of the East African population thought to breed in any one year.

Marabou storks at the carcass of an African elephant in Zimbabwe. Marabous dominate vultures when fighting over carrion.

Protective parents

At the beginning of the breeding season marabous gather in their nesting trees. The male can get very aggressive at first, inflating his throat pouch and driving off approaching females. Eventually the females are allowed to approach their prospective mates, and pairs are formed. The male then sets about collecting sticks, which the female weaves into a nest about 3 feet (90 cm) across, with a lining of twigs and green leaves. The nest is never left unattended as other marabous do not hesitate to steal sticks where they can.

Nest-building takes a week or more and sticks are added throughout the season. When the main construction is finished, the female lays two or three chalky white eggs. Both parents incubate them until they hatch, 29–31 days later. The newly hatched chicks are almost naked, and their parents brood them or shade them from the sun for 2 weeks. From then they continue to shelter the young in bad weather. One parent always stands guard until the chicks are quite large, to prevent neighboring marabous from removing the nest from under the chick.

A well-balanced diet

Marabou chicks are fed by both their parents. Food is usually dropped on the nest for the chicks to pick up, and water is regurgitated into their mouths. Young marabous are fed many frogs, tadpoles, fish and mice as well as scraps of meat and offal, and this is one reason why marabous nest within easy reach of stretches of open water. Sometimes flying termites and locusts are brought to the nest.

When they are about 13–15 weeks old, the chicks begin to fly from branch to branch of the nest tree, flying from the tree a week or so later. Sexual maturity is not reached until marabous are at least 4 years old. The oldest birds in the wild are thought to be 25 years old or more.

MARABOU

CLASS	**Aves**
ORDER	**Ciconiiformes**
FAMILY	**Ciconiidae**
GENUS AND SPECIES	***Leptoptilos crumeniferus***

ALTERNATIVE NAME
Marabou stork

WEIGHT
8¾–19½ lb. (4–8.9 kg)

LENGTH
Head to tail: 3⅗–5 ft. (1.1–1.5 m); wingspan: 7¼–9½ ft. (2.2–2.9 m)

DISTINCTIVE FEATURES
Massive gray or pinkish bill; naked, pinkish head with white ruff; bare, inflatable throat pouch; huge, broad wings; gray upperparts; off-white underparts

DIET
Mainly carrion and food discarded by humans; also fish, termites, locusts, frogs, lizards, snakes, rats, mice and flamingos

BREEDING
Age at first breeding: 4–5 years; breeding season: dry season; number of eggs: 2 or 3; incubation period: 29–31 days; fledging period: 13–17 weeks; breeding interval: more than 1 year

LIFE SPAN
Up to 25 years or more

HABITAT
Open, dry savanna; swamps; river margins; often around fishing villages

DISTRIBUTION
Sub-Saharan Africa from Senegal east to Eritrea and south to South Africa

STATUS
Common

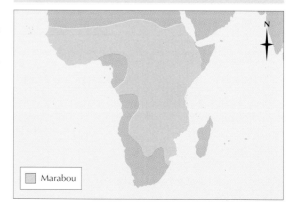

Marabou

MARINE IGUANA

THE MARINE IGUANA IS unique in its way of life, being the only truly marine lizard. It is found only on the larger islands of the Galapagos Archipelago, some 600 miles (965 km) west of Ecuador. It is of great scientific interest because of its exceptional home, and also its adaptations to feeding at sea.

The marine iguana is a heavily built lizard that grows up to about 4 feet (1.2 m) in length. It has a blunt snout, sturdy legs with partially webbed feet, and a crest of spines that runs from the neck to the tip of the tail. The tail is flattened sideways and is used for swimming. Most marine iguanas are black or very dark gray, but on Hood Island at the south of the Galapagos Archipelago their bodies are mottled with black, orange and red and their front legs and crests are green.

Rest up on lava fields

Marine iguanas are found on rocky seashores, entering the intertidal and subtidal zones to feed. Outside the breeding season, and when they are not feeding, they gather in tight groups on land, sometimes even piling on top of each other in heaps. They lie on the lava fields that are prominent features of the Galapagos. In the heat of the day they seek shelter under boulders, in crevices or in the shade of mangroves.

At the beginning of the breeding season, the males establish small territories, so small that one iguana may be on top of a boulder while another lies at the foot. Fights occasionally break out, but disputes are generally settled by displays. A male marine iguana threatens an intruder by raising itself on stiff legs and bobbing its head with mouth agape, showing a red lining. If this does not deter the other lizard, the owner of the territory advances and a butting match takes place. The two push with their bony heads until one gives way and retreats.

While marine iguanas are basking, large red crabs sometimes walk over them, pausing every now and then to pull at the iguanas' skin, removing the ticks that often infest these reptiles.

Diving for a living

As the tide goes down, the marine iguanas take to the water and eat the algae and seaweed exposed on the reefs and shores. They cling to the rocks with their sharp claws, so as not to be dislodged by the surf, and slowly work their way over the rocks. They tear strands of algae by gripping them in the sides of their mouths and twisting to wrench them off. At intervals they pause to swallow and rest. Marine iguanas sometimes swim out beyond the surf and dive to feed

A marine iguana resting on the rocky lava seashore. These animals are unique to the Galapagos Islands and are the only lizards that feed at sea.

Marine iguana feeding by tearing seaweed from underwater rocks. Its heart rate slows down while it is in the sea so that it does not lose heat too quickly.

MARINE IGUANA

CLASS	**Reptilia**
ORDER	**Squamata**
SUBORDER	**Sauria**
FAMILY	**Iguanidae**
GENUS AND SPECIES	***Amblyrhynchus cristatus***

ALTERNATIVE NAME
Galapagos marine iguana

LENGTH
Up to 4 ft. (1.2 m)

DISTINCTIVE FEATURES
Heavy body; blunt snout; partially webbed feet; flattened tail; crest of spines from neck to tip of tail; black to dark gray in color; some individuals mottled black, orange and red, with green crest and forelimbs

DIET
Seaweed and other marine algae

BREEDING
Number of young: 2 or 3; hatching period: about 110 days; breeding interval: 1 year

LIFE SPAN
Not known

HABITAT
Rocky seashores, entering intertidal and subtidal zones to feed

DISTRIBUTION
Larger islands of Galapagos Archipelago

STATUS
Locally abundant

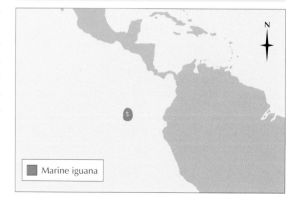

on the seabed. They have been recorded as feeding at depths of 35 feet (10.5 m) but will not usually dive below about 15 feet (4.5 m). The length of each dive is about 15–20 minutes.

Like all reptiles, marine iguanas must maintain a certain body temperature in order to remain active. When a marine iguana enters the cold sea water to feed, its body temperature might be expected to fall rapidly, quickly rendering the lizard helpless. However, the marine iguana is able to slow its heart rate down, so limiting blood circulation and reducing heat loss. When it comes back to shore its heart rate returns to normal and it rapidly warms up, basking in the sun.

Harems around the rocks

When the male iguanas have formed their territories, the females join them. They are free to move from one territory to another, but the males soon gather harems of females around them and mating takes place without interference from other males.

When the males leave their territories, the females gather at the nesting beaches. There is competition for nest sites and fighting breaks out. Each female digs a 2-foot (60-cm) tunnel in the sand, scraping with all four feet. She then lays just two or three white eggs in the tunnel. Each measures 3¼ by 1¾ inches (8 by 4.5 cm). Then the female iguana fills up and camouflages the tunnel. When the eggs hatch after about 110 days, 9-inch (23-cm) baby iguanas emerge.

Young preyed on by cats and dogs

Apart from humans, the main enemies of full grown marine iguanas are sharks, but the iguanas usually stay inshore where sharks are not likely to venture. Young iguanas are caught by herons, gulls and Galapagos hawks. Although these lizards are quite abundant in places, the eggs and juveniles are also vulnerable to predation by introduced animals such as cats, dogs, pigs and rats. Marine iguanas are strictly protected, and disturbance by increasing numbers of tourists is controlled.

MARINE TOAD

SUGARCANE HAS BEEN spread from an unknown native home in East Asia to warm countries around the world, including tropical America. With its spread, so the gray cane beetle, *Lepidoderma albohirtum*, has also been inadvertently introduced, the larvae of which live on the sugarcane roots and can destroy the plants. The marine toad feeds on cane beetles and, in an effort to control the beetle, it has been introduced around the world wherever sugar is grown. On the whole, this has failed to control the cane beetle, but the toad itself has successfully colonized many new areas.

Largest toad of all

Despite its name, the marine toad does not live in the sea, but is tolerant of brackish (slightly salty) water. Its habits are similar to those of more familiar toads, such as the common European toad, *Bufo bufo*. At first sight its main claim to fame is its size. It is the largest toad in the world and is sometimes called the giant toad. Usually the male is about 4 inches (10 cm) long and weighs ¾ pound (350 g). However, females are larger than males and may be 10 inches (25 cm) long and can weigh up to 3 pounds (1.4 kg). The marine toad is mottled yellow and brown and has many dark, reddish brown warts on its body.

The marine toad's native home is from southeastern Texas, through Central America to South America as far south as Patagonia. However, it has also been introduced into Florida, many Caribbean islands, Bermuda, Hawaii, New Guinea and Australia, and to a few other places where sugarcane is grown.

Prefers the damp

During hot weather the marine toad, also known as the cane toad in Australia, remains under cover of vegetation or burrows in the ground. It mainly comes out at night, or in wet or cool weather, to feed. It feeds on almost anything that moves that is small enough for it to swallow, especially insects and beetles. The marine toad also takes small vertebrates, such as smaller toads and frogs, small snakes and even mice, and sometimes eats vegetable matter.

Extremely rapid growth

Mating is normal for a toad, that is by amplexus (the mating embrace of a frog or toad). The eggs are laid into the water in very large clutches, perhaps 30,000 eggs or more per spawning. Each female lays several batches of eggs a year. Marine toads grow extremely rapidly after

completing the larval (tadpole) stage. The young toads are ⅓–½ inch (8–12 mm) long when they first emerge. They reach 2⅓–3 inches (6–7.5 cm) within just 3 months and 3½–5 inches (9–12 cm) within 6 months. Marine toads are sexually mature after 1 year.

Toxic toad

When it is attacked, the marine toad is one of the most poisonous of all toads. Its poison may cause closed eyes and a swollen face in humans, with other unpleasant symptoms such as nausea and vomiting. In extreme cases it may result in death, for the poison acts like digitalis and slows the heartbeat, so it can lead to heart failure. The poison is a whitish fluid exuded in small doses

The marine, giant or cane toad is the largest toad in the world. It is also well known for being highly toxic.

Amplexus is the mating embrace of a toad (or frog) during which the female sheds eggs to be fertilized by the male.

from the parotid glands (glands on each side of the head). The toad also inflates itself when disturbed, as many toads do, by gulping air to make it appear larger than it actually is.

In its native home the marine toad has few, if any, predators. In countries where it has been introduced, the native predators do not have the instinct to leave it alone, largely because it has no warning colors. Those animals that try to eat it are either poisoned or are suffocated when, having quickly gulped down the toad alive and whole, it inflates itself and blocks the throat or the gullet of the predator.

Destructive intruders

In 1863 sugarcane was planted in Queensland, Australia, near the town of Brisbane. Because of the success of this venture more land was cleared for sugar, and some of the native animals lost their habitat. Others, more suited to cultivated land, actually benefited. In 1935 the marine toad was imported from Hawaii, where it had been introduced in the hope it would eradicate the cane beetle. This was also the hope in Queensland, but the cane beetle still survives and the toad has become established and widespread.

The marine toad has had a drastic effect on the native wildlife. The smaller frogs and toads have suffered in numbers from being eaten by the marine toad, while snakes and birds such as ibises have been killed when attempting to eat the toad, either by being poisoned or suffocated. An unexpected damage is caused by the toad in dry summers when ponds and waterholes dry up. Such water that remains becomes so packed with breeding toads and their spawn that it is undrinkable. It is thought that the toad may spread throughout Australia in time. It is an adaptable animal, capable of fasting for up to 6 months.

MARINE TOAD

CLASS	**Amphibia**
ORDER	**Anura**
FAMILY	**Bufonidae**
GENUS AND SPECIES	***Bufo marinus***

ALTERNATIVE NAMES
Cane toad; giant toad

WEIGHT
¾–3 lb. (0.35–1.4 kg)

LENGTH
4–10 in. (10–25 cm)

DISTINCTIVE FEATURES
Extremely large size; mottled yellow and brown in color with dark reddish brown warts; exudes a milky white, highly toxic, secretion from parotid glands

DIET
Insects and small vertebrates, including other toad species; some vegetable matter

BREEDING
Age at first breeding: 1 year; number of eggs: more than 30,000 per spawning; hatching period: not known; breeding interval: several spawnings per year

LIFE SPAN
Not known

HABITAT
Savanna, marshes, woodland and forest; also suburban areas and cultivated land, especially where sugarcane is grown

DISTRIBUTION
Native range: southeastern Texas south to South America; introduced to Florida, the Caribbean, Bermuda, Puerto Rico, Hawaii, Australia, New Guinea and Philippines

STATUS
Common to abundant

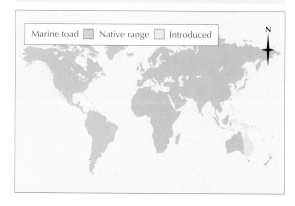

Marine toad ☐ Native range ☐ Introduced

MARKHOR

To many zoologists the markhor is the most striking and imposing of the wild goats. The male reaches 3 feet (90 cm) at the shoulder, occasionally up to 3¼ feet (1 m), and weighs 175–240 pounds (80–110 kg). The female is smaller, not quite as tall and less than half this weight. The male has prominent, thin, spiral horns, flattened and ribbonlike in some animals, keeled both in front and behind, and divergent. The straight length of these horns is 1⅔ to 3⅓ feet (0.5–1 m). Taken along the spiral, they are 2½–5½ feet (0.75–1.65 m) long, and the tip-to-tip distance ranges from ⅔ to 3⅔ feet (0.2–1.1 m). Females' horns are not more than 10 inches (25 cm) long.

The markhor is colored light reddish to sandy in summer, gray in winter, with a white to gray belly and a blackish tail. The male has a mane on the throat and breast that reaches 1 foot (30 cm) long in winter, and a mane on the neck and back that is half as long. He also has a prominent dark beard that reaches 8 inches (20 cm).

The markhor lives in the precipitous mountains of Central Asia. These include the Hindu Kush mountain range in Afghanistan and its outliers in Turkmenistan, Uzbekistan and Tajikistan, the Sulaiman ranges in Pakistan and the Astor, Gilgit, Pir Panjal and Chitral ranges in Kashmir.

Goats of the lower heights

Markhors live on the slopes of deep rocky gorges, where their agility and surefootedness stand them in good stead. The markhor is not found at such high altitudes as the ibexes. In Kashmir, for example, there are Siberian ibexes as well as markhor. Although the two species live in the same ranges, markhors tend to keep to lower altitudes. Generally they are found between 4,500 and 8,000 feet (1,400–2,400 m). In summer some individuals, mainly adult males, go higher into the true alpine zone, at about 10,000–12,000 feet (3,000–3,600 m).

Small, single sex herds

For most of the year, markhor herds are very small, seldom more than five animals, and consist entirely of either males or females. In the breeding season, around November to December in the Hindu Kush, herds of up to 27 are formed.

Like other goats, the males fight during the breeding season, rearing up and clashing their horns together. Gestation is 140–170 days, so the kids (young) are born at the end of April or the beginning of May. The female markhor leaves her group to give birth and then rejoins it when the kid can travel and keep up with the adults.

Predators and hunting

The markhor's range is also the home of the snow leopard, *Panthera uncia*. This large cat undoubtedly preys upon the markhor, although it is now itself endangered and unlikely to have a large impact on numbers. At somewhat lower altitudes the true leopard, *P. pardus*, wolves and probably bears also make inroads into the population. Raptors (birds of prey), foxes and perhaps the manul or Pallas's cat, *Felis manul*, take kids.

Finally, the human element must not be discounted. The markhor has been hunted extensively over the years for its horns and it is this that has been the main cause of the animal being endangered. The markhor's range is now considerably reduced and there are thought to be less than 5,000 individuals remaining worldwide.

The markhor, seen here in Afghanistan, used to range widely through the mountains of Central Asia. Hunted extensively for its horns, this wild goat is now endangered, with few animals remaining in the wild.

The male markhor is distinctive for its long, corkscrew-shaped horns and heavy mane on the throat and chest.

MARKHOR

CLASS	**Mammalia**
ORDER	**Artiodactyla**
FAMILY	**Bovidae**
GENUS AND SPECIES	*Capra falconeri*

WEIGHT
Male: 175–240 lb. (80–110 kg); female: 70–110 lb. (32–50 kg)

LENGTH
Head and body: 4½–6 ft. (1.4–1.8 m); shoulder height: 2–3¼ ft. (0.6–1m); tail: 3–5½ in. (8–14 cm)

DISTINCTIVE FEATURES
Large, stocky wild goat; reddish to sandy coat in summer, gray in winter; pale belly; blackish tail. Male: prominent dark beard; prominent, thin, twisted and upright horns. Female: shorter horns.

DIET
Mountain grasses and herbs; also leaves

BREEDING
Age at first breeding: 2–3 years (female); breeding season: November–December in many parts of range; number of young: 1 or 2; gestation period: 140–170 days; breeding interval: 1 year

LIFE SPAN
Up to 12 years

HABITAT
High mountains above tree line but not highest mountain tops

DISTRIBUTION
Central Asia: Turkmenistan, Uzbekistan, Tajikistan, Afghanistan, northern Pakistan and Indian Kashmir

STATUS
Endangered; estimated population: no more than about 5,000

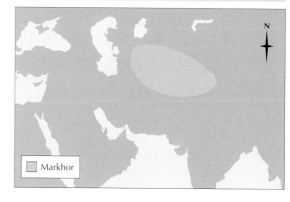

Markhor

Many twists and turns

Because it lives high up in the inaccessible mountains, the markhor is not among the best-known of the family Bovidae. Although it is so distinctive, the species was not made known to Western science until 1839. Its common name is from the Persian *mar* (snake) and *khor* (killer), but it is not known whether or not it kills snakes.

Numerous subspecies or races have been named, based almost entirely on the markhor's horns, the shape of which can vary enormously. The race inhabiting the Astor mountain range tends to have very divergent horns, each forming an open spiral of one or two complete turns. At the other extreme is the markhor of the Sulaiman range, which has horns that form a tight "V," with each a close corkscrewlike spiral of three or four complete turns. The more open type of horns predominate in Kashmir and northern parts of the markhor's range, with the tight "Vs" in the south.

MARLIN

THE STRONG AND SWIFT MARLIN is one of the most popular game fish. Its body is long and flattened from side to side, and the snout and upper jaw are drawn into a slender bill that is round in cross section. The name marlin derives from the word "marlinespike," a tool that tapers to a sharp point and is used to separate strands of wire or rope.

The marlin's dorsal fin is low and continuous in young fish, but with age the front part increases in height and the first few spines become greatly thickened. The anal fin is divided into two parts, and the pelvic fins are longer than the pectorals in the young fish but become relatively shorter with age. The tail fin is strongly forked, and there are keels at the base of the tail fin. The back is bluish, dark brown or black, and the underside is silvery, silver gray or yellow. In some species the back and flanks are marked with narrow blue or silver bands.

There are three species of true marlins belonging to the genus *Makaira*. The six related species of spearfish, some of which are also referred to as marlins, belong to the genus *Tetrapturus*. The Atlantic white marlin, *T. albidus*, grows to more than 9¾ feet (3 m) long and 183 pounds (83 kg) in weight. The blue marlin (*M. mazara*), striped marlin (*T. audax*) and black marlin (*M. indica*) are all native to the Indian Ocean and South Pacific. All three species grow up to 16 feet (5 m) in length and may weigh more than 1,800 pounds (820 kg). The shortbill spearfish, *T. angustirostris*, grows up to 7½ feet (2.3 m) long and 115 pounds (52 kg) in weight, while the longbill spearfish, *T. pfluegeri*, may reach 8⅓ feet (2.5 m) in length and weigh 100 pounds (45 kg). Along with swordfish and sailfish, marlins and spearfish are sometimes known as billfish, a reference to their long bill or snout. The marlin's snout is also known as a rostrum.

Ocean sprinters

Marlins are very powerful fish and are probably the fastest of all swimmers. They can reach speeds of 40–50 miles per hour (64–80 km/h) and perhaps even more. Such speeds are possible because of their shape; the body is streamlined, with the bill forming a highly efficient cutwater.

When marlins are going at full speed all of their fins, apart from the tail fin, are folded down into grooves in the body so there are no obstructions to an easy passage through water. Having the pelvic fins far forward, on a level with the pectoral fins, means that marlins and spearfish can turn suddenly in a tight circle.

Rapid movement through water places a great strain on the skeleton, especially when the fish has to brake suddenly. The marlin's backbone is made up of relatively few vertebrae, and each of these has flattened interlocking processes that give strength and rigidity to the whole body. Some indication of the speed and thrust of a marlin is seen in the way the bill has at times been driven through the timbers of ships. It seems likely that such incidents are due to accidental collision rather than deliberate attack.

Marlins lead solitary lives, well scattered about the ocean. In spring and summer they form pairs, suggesting that this period probably marks their mating and spawning season.

Relentless hunters

Marlins and spearfish feed mainly on other fish, especially mackerel and flying fish, although they will also take squid and cuttlefish. They pursue the shoals for days on end, striking to left and right with the bill and then feeding at leisure on the dead and injured victims. It seems that the billfish pursue, then stop to feed, overtake once more to attack their prey, stop again to feed, and

The marlin's powerful, streamlined body and superb maneuverability make it probably the fastest of all fish.

In common with other marlin species, juvenile blue marlin lack the elongated snout that characterizes the adult fish.

ATLANTIC BLUE MARLIN

CLASS	**Osteichthyes**
ORDER	**Perciformes**
FAMILY	**Istiophoridae**
GENUS AND SPECIES	***Makaira nigricans***

WEIGHT
Male: up to 330 lb. (150 kg); female: up to 1,810 lb. (820 kg)

LENGTH
Male: up to 13¾ ft. (4.2 m); female: up to 16½ ft. (5 m)

DISTINCTIVE FEATURES
Long, sharp snout; anterior tip of dorsal fin pointed or sharply rounded, rest of fin low; slim pelvic fins; bluish or dark gray blue above; silvery below

DIET
Fish, squid and cuttlefish

BREEDING
Breeding season: mainly July–September

LIFE SPAN
Probably up to 30 years

HABITAT
Offshore surface waters; coastal areas with adjacent deep water

DISTRIBUTION
Tropical and temperate Atlantic

STATUS
Common

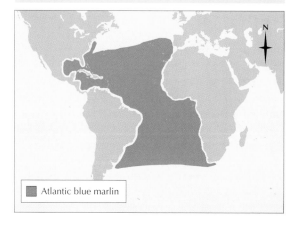

Atlantic blue marlin

so on. A fish that moves at great speed needs a large amount of food to supply the necessary energy to maintain such a demanding lifestyle.

The fish's strenuous muscular action generates a certain amount of heat. Tests have revealed that the temperature of a striped marlin, about 9 feet (2.7 m) long and weighing nearly 300 pounds (136 kg), was up to 11°F (6°C) higher than that of the surrounding water. A number of marlins were tested with a thermopile harpoon—a harpoon carrying a device for registering temperature. They were played for a half hour or more before being landed. This event always creates an increase in the fish's body temperature, but the test results suggested that the act of swimming at speed will cause any billfish to temporarily raise its blood temperature.

Vulnerable when exhausted?

The only natural predators of billfish are large sharks, especially tiger and great white sharks. Sharks are not fast swimmers. The blue shark has been estimated to reach 26 miles per hour (42 km/h) and the mako shark, which may attain speeds of 35 miles per hour (56 km/h), is probably as fast as any shark. It is possible that, like the cheetah, the fastest animal on land, marlins and spearfish can maintain high speeds for only short distances. The need to recover after such exertions may leave the fish vulnerable to the slower but persistent sharks.

Spearing their food

In the past there was some debate between deep-sea anglers as to whether marlins and other billfish ever spear their prey using their elongated snouts. In 1955 the *John R. Manning*, a longline ship of the U.S. Fish and Wildlife Service, captured a marlin in the seas south of Hawaii. It weighed 1,500 pounds (680 kg) and in its stomach was a freshly dead yellowfin tuna, 5 feet (1.5 m) long and weighing 157 pounds (71 kg). The tuna had been swallowed headfirst, and had two holes through the body that corresponded to wounds from the marlin's snout.

MARMOSET

MARMOSETS, THE SMALLEST of all monkeys, are found in South America. They differ from the more familiar monkeys in having claws and in lacking wisdom teeth, the third molars. These characteristics, along with the brain, are thought to be primitive features. For these and other reasons the marmosets are put, together with the tamarins, in the family Callitrichidae, separate from other South American monkeys.

Marmosets are arboreal (tree-living) and squirrel-like in appearance, often with dense, silky fur, distinctive ear tufts and a long, bushy tail. Their lower incisors are almost as long as the canines and are used in grooming. They have distinctive, twittering voices.

Smallest monkey in the world

There are 18 species of marmosets, all but one belonging to the genus *Callithrix*. Most are about 7 inches to 1 foot (18–30 cm) in head and body length, with tails of about ½–1⅓ feet (17–40 cm) in length. They weigh in the region of ½–1 pound (230–450 g). Smaller than this is the single species of the genus *Cebuella*, the pygmy marmoset, *C. pygmaea*. It is under 6 inches (15 cm) long, with a 7–9-inch (17–23-cm) tail. Weighing about 3–5 ounces (85–140 g), it is the smallest monkey in the world.

A variety of colors

Marmosets are found in the tropical and subtropical forests of Brazil, Bolivia and Paraguay. They vary in color depending on species, from white and silver to golden yellow, reddish and black. The white ear-tufted marmoset, *Callithrix jacchus*, from the Brazilian coast, for example, has fur marbled with black and gray. It has a black head and, as the name suggests, long, white tufts of hair around the ears. As with most marmosets, the male and female look much the same. The silky marmoset, *C. chrysoleuca*, is pale with silvery fur, while Geoffroy's marmoset, *C. geoffroyi*, has a pale head with a dark body and dark ear tufts. The pygmy marmoset, found in western Brazil, Colombia, Ecuador, Peru and Bolivia, has no ear tufts. The hair on its head is swept backward over its ears and it is brown in color, marbled with tawny.

Small family groups

Marmosets are extremely active animals and move about mainly during the day. With the aid of claws on all digits except the big toe, they bound and scamper along the branches of trees, often with quick, jerky movements. They live in small, family groups or troops of between four and 15 individuals. These are usually made up of a mated pair and various generations of their offspring. Even after reaching sexual maturity, young marmosets will often continue to live in the same family group. Marmosets live in the lower levels of the forest canopy, up to 62 feet

A young silvery marmoset, Callithrix argentata. Young marmosets will often continue to live in the same family group, even after reaching sexual maturity.

MARMOSETS

CLASS	**Mammalia**
ORDER	**Primates**
FAMILY	**Callitrichidae**
GENUS	**Marmosets, *Callithrix*; pygmy marmoset, *Cebuella***
SPECIES	**17 species in *Callithrix*. Single species in *Cebuella*: pygmy marmoset, *C. pygmaea*.**

WEIGHT
***Callithrix* species: ½–1 lb. (230–450 g).
Pygmy marmoset: 3–5 oz. (85–140 g).**

LENGTH
***Callithrix* species. Head and body: ½–1 ft.
(18–30 cm); tail: ½–1⅓ ft. (17–40 cm).
Pygmy marmoset. Head and body: 4¾–6 in.
(12–15 cm); tail: 7–9 in. (17–23 cm).**

DISTINCTIVE FEATURES
**Small size; distinctive ear tufts; long, bushy
tail; color varies between species**

DIET
**Fruits, leaves, nuts, tree sap, invertebrates,
small vertebrates and eggs**

BREEDING
**Age at first breeding: not known; breeding
season: all year; number of young: 1 to 3,
twins common; gestation period: 140–145
days; breeding interval: not known**

LIFE SPAN
Up to 10 years

HABITAT
Tropical and subtropical forest

DISTRIBUTION
South America

STATUS
**Some species and subspecies endangered,
several others vulnerable or threatened**

Marmosets

The pygmy marmoset belongs to a different genus than other marmoset species. Much smaller than the other 17 marmosets, it is the smallest monkey in the world.

(19 m), but sometimes go higher, or down to the ground in search of fallen fruits. They feed on invertebrates, small vertebrates, eggs, tree sap, fruits and leaves. The pygmy marmoset takes more fruits, nuts and tree sap than the others.

For a long time there was some doubt as to whether the family groups were territorial, because studies were largely made of captive animals. It has since been found that in the wild marmoset troops occupy overlapping ranges. When marmosets from separate troops meet, they threaten each other with a rapid flattening and erection of the conspicuous white ear tufts.

Attentive fathers

Breeding may take place at any time of the year. Unlike many other higher primates, marmosets have a courtship display. The male walks with his body arched, smacking his lips and pushing his tongue in and out. The male and female lick each other's fur and groom each other using their long, lower incisors as a comb. As in other monkeys, the female enters estrus (comes into season) at roughly monthly intervals. At this time the male is very active. The gestation period is about 140 days and between one and four young are born. However, twins are the rule rather than the exception among marmosets. In 50 percent of the births of white ear-tufted marmosets, and in two-thirds of those of pygmy marmosets, there are twins. Triplet births are more common than single births.

The male assists at the birth, receiving and washing the newly born young. The male also carries the babies about, the young marmosets clinging to his back or around his neck. The female may carry them, but as a rule the father transfers the babies to the mother only at feeding time. Young marmosets are completely independent at 5 months.

Midday rest

Some tree-living animals, such as squirrels, have a group of long bristles on each wrist. These are tactile hairs (organs of touch), and marmosets are the only members of the higher primates to have this primitive characteristic. An even more primitive feature is the varying body temperature of the marmosets. This may vary by as much as 7° F (4° C), from one part of the day to another, and is lowest around midday. This suggests that the marmosets have a period of torpor (a state of inactivity almost like hibernation) at the time when most other mammals rest because the sun is at its hottest.

Threats to survival

There are several small cats as well as some raptors (birds of prey) in the Brazilian jungle that are thought to prey upon marmosets.

Marmosets have also been popular as pets since the early 17th century. At one time it was even fashionable for women to carry them inside their sleeves. In addition, marmosets used to be exported on a massive scale for use as laboratory animals in medical research.

These are important historical factors in the endangered status of many species today. However, the main threat to the marmosets, as with so many animals, is now loss of habitat. Logging and slash-and-burn agricultural techniques not only reduce valuable habitat but also destroy thousands of plant species, many of

which are relied on for food. The rapid shrinking of South America's rain forests has meant that a number of forms, such as the buffy tufted-ear marmoset, *C. aurita,* and the buffy-headed marmoset, *C. flaviceps caroli,* are now endangered. Only a few thousand buffy-headed marmosets remain and their future, in particular, looks bleak. Several others are vulnerable or threatened, including Geoffroy's marmoset, the silky marmoset and the white marmoset, *C. argentata leucippe.* It can only be hoped that tourism in the region will justify the preservation of enough of the Amazon rain forest to prevent the marmosets' habitat being destroyed completely.

An emperor marmoset, Callithrix imperator.
A number of marmoset species are endangered or threatened. Although they are no longer exported for use as laboratory animals, loss of habitat is now the major problem facing these monkeys.

MARMOSET RAT

rats can grasp bamboo stems by gripping with both first and fifth toes opposed to the three central toes.

Sure-footed climbers

Marmoset rats can climb the slippery surfaces of the stout stems of bamboo by spreading their toes wide. They climb steadily and without difficulty, the claws taking no part in the action. The rats can also stop and stand or turn at any angle. On a stem leaning from the vertical they grip with the soles of the feet opposed, as a human climber might grip a pole, and move in short spurts or bounds, alternately moving both forefeet and then both hind feet.

The naked skin on the undersides of the feet also helps in climbing. As with all rats, there are pads on the soles and narrow transverse ridges on the undersides of the toes. While the marmoset rats are active, the skin on the toe-pads is moistened by a sticky secretion that gives extra grip.

Bamboo nests

Marmoset rats enter the stems of a bamboo to nest by gnawing a hole about 1¾ inches (4.5 cm) in diameter. Once inside they may use only the one cavity between the divisions across the stem, or they may gnaw through them to occupy several cavities. At times the rats make an exit hole, gnawed through from the inside. The nests themselves are made exclusively of bamboo leaves.

The stem structure of *Gigantochloa scortrchinii* seems to be the reason for the rats' choice of this bamboo species. In other species of bamboo growing in the same area, the walls of the stems are thin and the fibers are concentrated at the surface. This not only makes the surface difficult to penetrate but also gives the walls a tendency to split longitudinally, and they break easily under stress when holes are cut in them.

Meals of bamboo

The diet of a captive marmoset rat was once studied. It was offered fruit to supplement its diet and it took a little papaya, banana and sweet potato. However, when it had learned to drink from a water bottle, it rejected the fruit, suggesting that it only took this food for the water it contained. The rat would also take paddy (unmilled rice), but it obviously preferred the

Dependent on a single species of bamboo, marmoset rats are a good example of highly specialized rodents. They are quite unlike more adaptable species such as the brown rat, Rattus norvegicus.

ALTHOUGH NEVER CALLED bamboo rats, the marmoset rats' very existence depends on a single species of bamboo, *Gigantochloa scortrchinii*, found in Southeast Asia. There are two species of marmoset rats, *Hapalomys longicaudatus* and *H. delacouri*, both of which are also called Asiatic climbing rats. They are found in tropical forests and bamboo stands in parts of Peninsular Malaysia, Thailand, Laos, Vietnam and Myanmar (Burma).

Marmoset rats are stocky rodents, 4¾–6⅔ inches (12–17 cm) in head and body length with long, 5½–8-inch (14–20-cm) tails. The plump body is covered with a soft, silky fur, which is grayish brown turning to reddish or chestnut red with white underparts. The coat molts back to grayish brown on top. The eyes are large and black. The ears are also large and the nose is blunt. The scaly tail has scanty hairs except in the final third, which is hairy and looks like a bottlebrush.

The rats' feet are adapted for gripping bamboo stems. They are short and broad, pink and coated with white hair. Each toe, which is long and broad-ended, has a sharp claw at the tip except for the first toe of the hind foot, which has a small nail. This toe is opposable to the rest. The

growing tips of bamboo twigs, and especially fruiting twigs. When deprived of bamboo it was found to rapidly lose weight and condition.

From another line of investigation it seems that the marmoset rats' diet is even more exclusive. Examination of the stomach contents of rats taken in the wild showed large quantities of pollen. While there could not be absolute proof of this, the pollen was probably bamboo and almost certainly that of the one species of bamboo. Many species of bamboo in Southeast Asia flower rarely, some only at intervals of several years. *G. scortrchinii*, however, flowers, and therefore produces pollen, frequently.

MARMOSET RATS

CLASS	**Mammalia**
ORDER	**Rodentia**
FAMILY	**Muridae**
GENUS AND SPECIES	***Hapalomys delacouri;*** ***H. longicaudatus***

ALTERNATIVE NAMES
Asiatic climbing rat; marmoset mouse

LENGTH
Head and body: 4¾–6⅔ in. (12–17 cm); tail: 5½–8 in. (14–20 cm)

DISTINCTIVE FEATURES
Stocky body; long, scaly tail with hair on final third; short, broad feet covered with white hair; large, naked toes with sharp claws; soft, silky fur; grayish brown to chestnut upperparts; whitish underparts

DIET
Shoots, pollen and fruits from single species of bamboo, *Gigantochloa scortrchinii*

BREEDING
Poorly known

LIFE SPAN
Not known

HABITAT
Tropical forests and stands of bamboo

DISTRIBUTION
Southeast Asia: Myanmar (Burma), Laos, Vietnam, Thailand and Peninsular Malaysia

STATUS
Both species: near threatened

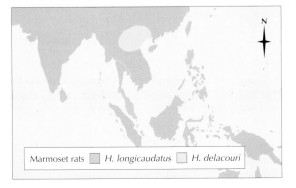

Marmoset rats ▢ *H. longicaudatus* ▢ *H. delacouri*

History of the rats' discovery

Nothing is known of the marmoset rats' breeding habits or of their predators. In fact until recently they were almost unknown animals. At first they were known as Berdmore's rats, after Major Berdmore, who in 1859 collected a single specimen from Tenasserim in southeastern Myanmar. At this time the species was given the scientific name *Hapalomys longicaudatus*, meaning the long-tailed soft mouse. In 1915 another specimen was taken in Peninsular Malaysia, and in 1927 a third was found in Laos. In the same year a specimen collected in Hainan, an island in the South China Sea, was described by the American zoologist A. M. Allen and given the colloquial name of marmoset mouse. In 1956 another was trapped at Selangor in Peninsular Malaysia, but so few specimens taken over such a wide area of Southeast Asia suggested rarity.

Then Lord Medway published his account of the marmoset rat, as he preferred to call it, in the *Malayan Nature Journal*, for August 1963. In January of that year he had found an area in Ulu Kelanton district where marmoset rats were common and well known. In fact, he found that the local Temiar people regularly ate these rodents and were skillful at hunting them. Since Medway's discovery zoologists have identified a second species of marmoset rat, which has been given the scientific name *H. delacouri*.

Threatened by bamboo destruction

The most notable feature of the marmoset rats is probably their close connection with a single bamboo species, and their adaptations to both a bamboo habitat and diet. However, intense specialization such as this leads to insecurity for a species. If anything occurs to upset the environment, the animal is unable to adapt readily to the changed circumstances. The marmoset rats offer us a clear example of the truth of this. If in any region the bamboo *Gigantochloa* is cleared, or for any reason it dies out, the marmoset rats disappear with it, being unable to adapt to other food sources. Although not as rare as was once thought, marmoset rats are now in decline across much of their range due to widespread human destruction of bamboo stands. Both species are now considered to be near threatened as a result, and may soon be classed as vulnerable.

MARMOT

Yellow-bellied marmots, M. flaviventris, sparring. Found in high, rocky parts of the western United States and British Columbia, this species is sometimes an agricultural pest.

THERE ARE 14 SPECIES OF marmots. Among the best-known of these are the alpine marmot, *Marmota marmota*, and the bobak, *M. bobak*, both of Europe, and the hoary marmot, *M. caligata*, of North America. There are various other alpine species in both the Old and the New World. Equally well-known is the groundhog or woodchuck, *M. monax*, of North America. This marmot differs markedly in its habits and is discussed elsewhere.

Marmots are large rodents, ranging in size from 1 to 2 feet (30–60 cm) long with a 4–10-inch (10–25-cm) bushy tail. They weigh 6½–15½ pounds (3–7 kg). Stout-bodied with short legs, the head is wide and short with small, rounded ears, large eyes and large, prominent teeth. The fur is thick, coarse and stiff and varies in color between species and seasons. Marmots live in underground burrows and are well-adapted for digging with their strong feet and claws.

Wide ranging rodents

The alpine marmot is a gregarious animal, living in colonies in and around the forest edges. It is found from about 4,000 to 9,000 feet (1,000–3,000 m) in the Alps and Carpathians and in corresponding alpine districts of central and northeastern Europe. The bobak, also called the Himalayan marmot or steppe marmot, is similar to the alpine species in many respects, but lives more in virgin grasslands. Because agriculture has taken much of its habitat, the bobak has become extinct over large areas of its former range, which used to be throughout Central and southern Asia and extreme eastern Europe. It is still found in parts of Central Asia and eastern Europe. The bobak is golden brown in color, with a black tail tip. The alpine marmot, meanwhile, is pale brown, its back and crown being peppered with black. It has white markings on its face and the outer half of its tail is black.

MARMOTS

CLASS	**Mammalia**
ORDER	**Rodentia**
FAMILY	**Sciuridae**

GENUS AND SPECIES **14 species, including bobak, *Marmota bobak*; alpine marmot, *M. marmota*; and hoary marmot, *M. caligata***

ALTERNATIVE NAMES
Bobak: Himalayan marmot, steppe marmot; hoary marmot: whistler

WEIGHT
6½–15½ lb. (3–7 kg)

LENGTH
Head and body: 1–2 ft. (30–60 cm); tail: 4–10 in. (10–25 cm)

DISTINCTIVE FEATURES
Very large, stocky rodent; thick fur; often gray, beige or orangish in color; long, bushy tail; small ears; large, prominent teeth

DIET
Grasses, sedges, low-growing plants and tubers; also fruits, nuts and invertebrates

BREEDING
Age at first breeding: 2 years; breeding season: spring; number of young: 4 or 5; gestation period: about 30 days; breeding interval: 1 year

LIFE SPAN
Usually up to 15 years

HABITAT
Grassland, steppe and alpine meadows

DISTRIBUTION
Bobak: Central Asia. Alpine marmot: central Europe. Hoary marmot: western North America. Some other species very localized.

STATUS
Several species common; others declining. Endangered: Vancouver Island marmot, *Marmota vancouverensis*.

Hoary marmot ☐ Alpine marmot ■ Bobak ☐

The hoary marmot is silvery gray in color, peppered with black on the back and rump. Its face is black with white cheeks, and the forehead, lower legs and feet are black. This species lives in mountainous parts of western North America, from Alaska south to New Mexico, as well as in the mountains of Siberia. Its habits and life history are similar to those of the alpine marmot.

Other species have a far more localized range, for example the Vancouver Island marmot, *M. vancouverensis*, is found only on Vancouver Island, while the Olympic marmot, *M. olympus*, occurs only in Washington State.

Mountain shelters

Marmots are diurnal (day-active) and leave their burrows on sunny, boulder-strewn slopes, as the first rays of the rising sun fall on them. When an entrance is shaded there may be some delay, and individual marmots differ, some seeming more reluctant than others to leave their burrows. The

Marmots, such as this alpine marmot, are large, stocky rodents with thick fur and large, prominent teeth.

The Olympic marmot (pictured above) is one of those species with an extremely isolated population, being found only in Washington State.

sleeping nests of dry grass are deep in their burrows, which may be several yards long and have several entrances. Late in the fall marmots change the grass in their sleeping chambers before going into hibernation for the winter. The whole family sleeps together, with the burrow entrances blocked.

Marmots spend the first hour after dawn sunning and grooming themselves near the burrow entrances, after which they disperse to the feeding grounds. They feed on grasses, sedges and herbaceous plants as well as roots and tubers. Fruits, nuts and invertebrates are also taken. Unlike many of the related ground squirrels, marmots have no cheek pouches for carrying food. Feeding lasts for about 2 hours in the morning, after which the animals resume their grooming and sunning. There is a further feeding period in the 2 to 3 hours before sundown before the marmots retire for the night.

Slow developers

Marmots mate soon after they emerge from hibernation in the spring. The gestation period is about 30–32 days and there are usually four or five young in a litter, born in the early days of June. The babies first come out of the burrows in

mid-July. They stay with the parents until the following spring. The female has one litter each year. Compared with other rodents, the young marmots are slow to develop and are not fully grown until 2 years old. Alpine marmots have been known to live up to 20 years in captivity, although 13–15 years is more common.

Sounding the alert

Marmots are too large to be preyed upon by the smaller predators such as ermines and are mainly taken by foxes and eagles. However, they are alert to danger and extremely wary. Marmots are often seen sitting upright, keeping watch, and are famous for their shrill alarm whistle. Those species that live in colonies have been credited with posting sentinels to stand guard. At the first sign of danger their whistles send the other marmots in the group scurrying to their burrows. The hoary marmot is also known as the whistler because of its sharp call. It is hunted by humans for both its flesh and its fur.

Several marmot species are in decline, but more as a result of agricultural development and isolation of populations. The Vancouver Island marmot is now endangered, while survival of the bobak is dependent on conservation measures.

MARSUPIAL CAT

THERE ARE SIX SPECIES of marsupial cats, or quolls, that are native to Australia, Tasmania and New Guinea. They are not true cats but carnivorous marsupials, which, together with the Tasmanian devil, make up the subfamily Dasyuridae. At one time the marsupial cats and the Tasmanian devil were placed in a single genus *Dasyurus*, but they are now separated as they differ in many respects.

In form marsupial cats are more like a civet or a genet than a cat. Their legs are short and their tails long and fairly bushy. The coat is gray to dark brown or black with pale spots. The head of some native cats is almost mouselike, with a pointed muzzle and long whiskers. Marsupial cats are, in fact, closely related to the marsupial mice

Pale-spotted cats

The eastern native cat, *Dasyurus viverrinus*, or eastern quoll (an aboriginal name), is one of the best known of the marsupial cats. It is about the size of a domestic cat, and occurs in two color phases. Some eastern quolls are black with large white spots, and others are grayish brown with creamy white spots. This last variant is more common but both may occur in one litter. Other marsupial cat species have similar coloring. They range in size from about half that of the quoll to about twice this size. One of the smallest is the little northern native cat, *D. hallucatus*. It is 9½ inches to just over 1 foot (24–35 cm) in length with a 8¼-inch to 1-foot (21–30-cm) tail. The largest species is the large-spotted native cat, *D. maculatus*, or tiger cat. It is 1½–2½ feet (40–75 cm) in head and body length with a 1–2-foot (30–60-cm) tail. This species is also known as the spotted-tailed quoll. Although shy animals, marsupial cats can be extremely fierce if cornered or threatened.

Species under threat

Some marsupial cats were at one time common animals of the Australian countryside. However, a number of species are now extinct in many parts of their former ranges and all are considered vulnerable or near-threatened. Certain subspecies are endangered.

Geoffroy's native cat, *D. geoffroii*, for instance, is now very much restricted to inland areas. It seems to have adapted itself to the advance of civilization, and has been known to breed in the roofs of houses. However, marsupial cats may attack and kill hens and poultry, so Geoffroy's native cat, and other species, are killed whenever possible by poultry farmers. Habitat loss, predation by domestic dogs and competition for food from domestic cats have also contributed to the decline of most species.

The little northern native cat is still quite common, however. It lives in northern Australia in rocky places or on wooded plains, where it shelters in crevices or hollow logs. It is found in abandoned houses and even in wrecked ships. Two closely related species, the New Guinea native cat, *D. albopunctatus*, and the bronze native cat, *D. spartacus*, are found only in New Guinea.

Marsupial cats are mainly arboreal (tree-living), the exceptions being the eastern native cat and Geoffroy's native cat, although these species can also climb well. The tree-dwelling marsupial cats have roughened pads on their feet, giving them a good grip on tree trunks and branches. The large-spotted native cat is so at home in the trees that it can run out to the very tips of branches.

Stealthy predators

The marsupial cats hunt at night and feed on small animals, mainly insects, snails, lizards and mammals such as rats, mice and young rabbits. When living near rivers they are partial to fish. They also rob nests of eggs and young birds. By eating rodents and rabbits the marsupial cats are of direct benefit to farmers, making up for some of the damage they do to poultry.

A little northern native cat, sometimes called the northern quoll. The smallest of the six marsupial cats, it is still quite common in northern Australia.

The eastern native cat is one of the best-known marsupial cats. It is now almost restricted to the forests and open country of Tasmania.

MARSUPIAL CATS

CLASS	**Mammalia**
ORDER	**Dasyuromorphia**
FAMILY	**Dasyuridae**
GENUS	***Dasyurus***

SPECIES **6, including little northern native cat,** *Dasyurus hallucatus*; **eastern native cat,** *D. viverrinus*; **and large-spotted native cat,** *D. maculatus*

ALTERNATIVE NAMES
All species: native cat, quoll, dasyure. Large-spotted native cat: tiger cat.

WEIGHT
1¾–6⅔ lb. (0.8–3 kg)

LENGTH
Head and body: 9½–30 in. (24–75 cm); tail: 8¼–24 in. (21–60 cm)

DISTINCTIVE FEATURES
Pointed muzzle; large ears; large eyes; gray to dark brown or black coat with many pale spots; long, bushy tail

DIET
Small birds and mammals such as rats and mice; also reptiles, invertebrates and eggs

BREEDING
Age at first breeding: not known; breeding season: May or June; number of young: up to 24, only 6 to 8 survive; gestation period: 16–20 days; breeding interval: not known

LIFE SPAN
Up to 6 years in captivity

HABITAT
Forest, bush, scrub, open farmland, savanna and mountainous areas

DISTRIBUTION
Australia, Tasmania and New Guinea

STATUS
All species vulnerable or near threatened; some subspecies endangered

Marsupial cats

Marsupial cats stalk their prey cautiously. The eastern native cat, for example, slowly approaches its victim, then leaps and kills it with a single bite to the back of the neck. The bigger large-spotted native cat can kill baby wallabies and large birds, but mainly eats small mammals, birds and reptiles.

Large litters

Marsupial cats breed in about May or June, the mechanism of birth being similar to that of other marsupials. The newborn young are minute, only about ¼ inch (0.6 cm) long in the large-spotted native cat. The babies become attached to teats enclosed by a simple pouch, often little more than a fold of skin. When 4 weeks old, the young large-spotted native cats are still only 1½ inches (4 cm) long. In general the young of marsupial cats leave the teats after 7–9 weeks and become independent by about 18 weeks, when they are one-third grown.

The teats number between six and eight in the different species of marsupial cats, and usually there are the same number, or fewer, young. However, female eastern native cats have been recorded as bearing up to 24 young. As each female has 6 or possibly 8 teats, most of such a large litter will die.

MARSUPIAL FROG

THERE ARE AROUND 30 species of South American tree frogs in which the females, instead of staying with the eggs to guard them, as some frogs do, carry the eggs about with them in pouches. The first scientists to discover this in 1843 merely saw that the female had a pouch and gave the frogs the scientific name *Gastrotheca*. Translated literally, this means "stomach pouch," although the pouch is actually on the female's back.

Marsupial frogs are fairly ordinary tree frogs except that some species are very small, ½–1 inch (1.5–3 cm) in length. The largest species are only 4 inches (10 cm) long. Marsupial frogs have sucker pads on the tips of the toes, usual in tree frogs, and some have stouter legs than most tree frogs. They are colored mainly green with brown spots, blotches or stripes.

As adults, marsupial frogs are mainly arboreal (tree-living). However, the highland species dwell less in trees and more in damp areas. All are generalist insectivores. Because of their unusual reproductive mode, they do not have to go to water to spawn.

Marsupial frogs are found in both Central and South America, from Panama, Colombia and Venezuela south through eastern Brazil as far as the southeast of that country.

Female parental care

Most species of marsupial frogs have an annual reproductive cycle, with only one brood of young each year. They are known for the unusual way in which the female takes care of the young. This varies to some degree from one species to another. However, in all species the female has a dorsal pouch where the eggs are held during development. Marsupial frogs can be divided into two broad groups: those species that brood the eggs, releasing the newly hatched tadpoles, and those species that brood the young until metamorphosis is complete.

Brooding the young

The majority of marsupial frogs, around 20 or so species, fall into this second category. The eggs remain in the pouch until they hatch and the tadpoles stay there also, leaving the pouch only when they have changed into froglets. In the smallest of these species, each female lays only four to seven eggs, rich in yolk. The larger species of this group may lay 50 or more eggs.

When these frogs are pairing, the male clasps the female as usual, but he is slightly farther forward on her back. Just as she is about to lay, the female raises herself on her hind legs so her back tilts steeply downward toward her head.

A marsupial frog, Gastrotheca marsupiata, one of the 11 or so species that brood the eggs, releasing the newly hatched tadpoles into ponds.

A female pygmy marsupial frog, Flectonotus pygmaeus, in Venezuela. A small brood of eggs can be seen waiting to hatch inside the pouch on her back.

MARSUPIAL FROGS

CLASS	**Amphibia**
ORDER	**Anura**
FAMILY	**Hylidae**
GENUS	***Gastrotheca* and *Flectonotus***
SPECIES	**Around 30 species, including *Gastrotheca marsupiata*, *G. mertensi* and pygmy marsupial frog, *Flectonotus pygmaeus***

LENGTH
½–4 in. (1.5–10 cm)

DISTINCTIVE FEATURES
Small tree frogs; sucker pads on tips of toes; some species have stouter legs than other tree frogs; most species green in color with brown markings. Female: dorsal pouch where eggs or young are held during development.

DIET
Insects

BREEDING
Varies between species. Number of eggs: 4 to 200; breeding interval: 1 year. All exhibit female parental care: some species carry eggs and release young as largish tadpoles into ponds, other species brood young until metamorphosis is complete.

LIFE SPAN
Not known

HABITAT
Mainly arboreal (tree-living) in rain forests; highland species more in damp areas rather than in trees

DISTRIBUTION
Parts of South America, from Venezuela south to southeastern Brazil; also Panama

STATUS
Not known

Her cloaca, the opening through which the eggs are laid, is directed upward. As a result, the eggs roll down her back and into the pouch. The male fertilizes each egg as it goes. Once the eggs are all safely inside, the mouth of the pouch closes giving a lumpy appearance to the dorsal surface. When the froglets are ready to be released, the female lifts a hind toe over her back and pulls apart the edges of the slitlike opening to the pouch.

Launching the tadpoles

Around 11 species, usually the high-dwelling, montane marsupial frogs, fall into the second group, the female carrying the eggs in the dorsal pouch until hatching. She then releases the young as largish tadpoles into ponds. These species also have a different way of putting eggs in the pouch. They lay up to 200 eggs and as the eggs are released by the female the male fertilizes them. At the same time he uses his hind feet to push the eggs into the pouch.

The female marsupial frog can feel when an egg is about to burst and release the tadpole. At this point she goes to water and lowers herself in. Then she brings her hind legs over her back and puts the first toe of each foot into the opening of the pouch and pulls, so the mouth of the pouch opens wide to let the tadpoles out. They come out one at a time, at short intervals.

Inside the pouch the tadpoles are enclosed in the egg membrane. Their external gills are much enlarged to form a sort of placenta for breathing. Soon after the tadpoles leave the pouch, the gills are quickly absorbed. The hind legs appear several weeks after the tadpoles enter water, then the forelegs grow. Eventually they change into froglets and metamorphosis is complete.

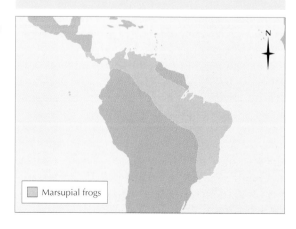

Marsupial frogs

MARSUPIAL MOLE

THE MARSUPIAL MOLE looks very like a true mole, except the female has the pouch of a marsupial for carrying her young. It used to be thought there were two species, but the current view is that there is only one.

The marsupial mole has a cylindrical body, short neck and legs and is covered with long, silky fur. The fur is pale but varies from silvery white to reddish or pink in color because of staining from the soil. It is 3½–7 inches (9–18 cm) in head and body length. There is a horny shield covering the snout. The only sign of an ear is a tiny hole covered with fur on each side of the head. The eyes are also vestigial (imperfectly developed). They are covered by fur and have no lens or pupil and only a weak optic nerve. Each forefoot has a cloven scoop formed by the curved and much enlarged claws on the third and fourth toes, the other three toes on the forefeet being small. The middle three toes of the hind feet also have slightly enlarged claws. The hard, leathery tail, ½–1 inch (1–2.5 cm) long, is marked with rings and knobbed at the tip.

The size of the marsupial mole seems to vary from one part of the animal's range to another. It is smallest in northwestern Australia. The mole is found in hot, sandy deserts, especially river flats, in south-central to northwestern Australia.

No permanent tunnels

The marsupial mole is a relatively recent discovery, for the non-indigenous inhabitants of Australia at any rate. The first one was not seen by Europeans until 1888. This is because it does not make permanent burrows, so it is far more difficult to find than the true moles with their distinctive molehills and permanent underground systems. This lack of tunnels also means that the marsupial mole has to surface regularly for air. After rain the mole will often come up onto the surface. Moving over the sand it leaves a characteristic triple track, made by the two forepaws and the stumpy tail. This track sometimes leads a naturalist to a marsupial mole before it dives back down into the sand.

These moles are thought to have been fairly common until recently, but numbers are now declining rapidly as a result of habitat loss.

Restless burrowers

Marsupial moles have the same restless, highly active way of life as true moles. They burrow about 3 inches (8 cm) below the surface, continuously hunting for food. In captivity marsupial moles have been found to eat earthworms and insects. Their diet in the wild is thought to comprise insect larvae, adult insects such as ants, and some plant material, for example seeds.

Feverishly active one moment, marsupial moles suddenly fall asleep, often after eating. When they wake up it is equally abrupt. They start their relentless foraging again, as if trying to crowd as much eating into the time as possible.

Life history little known

Nothing is known about the marsupial mole's breeding habits, apart from the fact that the female has a pouch for her young. Nor is it known how long this mole lives or if it is preyed upon by any other animals in the desert. As a rule, marsupial moles are brought to naturalists by Australia's indigenous people, the Aborigines. Several moles have been kept in captivity but their life history is still little understood.

The marsupial mole leaves no permanent tunnels as it burrows beneath the sand. As a result it has to surface regularly for air.

CLASS	**Mammalia**
ORDER	**Notoryctemorphia**
FAMILY	**Notoryctidae**
GENUS AND SPECIES	***Notoryctes typhlops***

WEIGHT
1½–2½ oz. (40–70 g)

LENGTH
**Head and body: 3½–7 in. (9–18 cm);
tail: ½–1 in. (1.2–2.5 cm)**

DISTINCTIVE FEATURES
**Resembles true, insectivorous moles.
Enlarged claws, especially on front paws,
forming 2 cloven scoops; horny shield on
nose; vestigial (imperfectly formed) eyes
and ears; short, ringed tail; long, silky fur;
pale silvery white to yellowish red, reddish
or pink in color.**

DIET
**Mainly insect larvae and adult insects such
as ants; some seeds and other plant material**

BREEDING
No details known

LIFE SPAN
Not known

HABITAT
Hot deserts, especially dry river flats

DISTRIBUTION
Western and south-central Australia

STATUS
Endangered

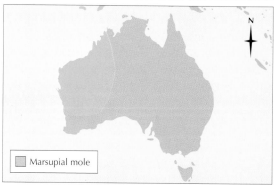

Marsupial mole

*When a marsupial
mole moves across the
sand, as it sometimes
does after rain, it
leaves a characteristic
triple track. This is
made by its two front
paws and by its tail
trailing behind.*

Evolutionist's dream

We are told that the first marsupial mole seen by
scientists caused as much excitement as did the
first platypus. By 1888 the furor caused by the
publication of Darwin's *Origin of Species* in
1859 had hardly abated. Scientists were still
searching for clues to support the general idea of
evolution when suddenly they were presented
with this first-class example of convergent or
parallel evolution.

Except for the color of its fur, the first marsupial mole looked almost exactly like the true
moles of the family Talpidae. Yet it was known
that it could not be related to the true moles

because of its pouch and geographical location,
poles apart from the other moles. It was therefore
understood that this similarity in appearance
must have evolved because the marsupial mole
lived in the same way and had the same habits
as the true moles.

MARSUPIAL MOUSE

MARSUPIAL MICE BELONG to the same family (Dasyuridae) as the Australian marsupial predators such as the native cats or quolls and the Tasmanian devil. These larger animals have a reputation for being voracious hunters, a quality shared by marsupial mice, if on something of a smaller scale.

There are around 55 mouse- or shrew-sized marsupials, variously known as marsupial mice or marsupial jerboas, depending on whether or not they have long hind legs. All but two species are found in Australia. They are delicately built with sharply pointed snouts, conspicuous vibrissae (whiskers) and large eyes and ears. In most species the hind legs are not markedly longer than the forelegs. Their fur is reddish brown to gray in color with white underparts. Marsupial mice can be distinguished from other rodents because instead of two incisors in the upper and lower jaws, they have eight incisors in the upper jaw and six in the lower jaw.

Some species, such as the broad-footed marsupial mice (genus *Antechinus*), range throughout Australia in suitable habitats, while others tend to be isolated to a particular region.

Most species in scrub and desert

There are several different groups of marsupial mice, each with its own characteristics reflected in its common name. There are the short-haired marsupial mice, genus *Murexia*, for example, and the brush-tailed marsupial mice, or tuans, of the genus *Phascogale*. Those belonging to the genus *Planigale* are often called flat-skulled marsupial mice, or planigales. There is also the fat-tailed dunnart, *Sminthopsis crassicaudata*, while the marsupial jerboas (genus *Antechinomys*) are obviously reminiscent of jerboas.

Marsupial mice can grow more than 1 foot (30 cm) long, of which slightly less than half is tail. Most are nocturnal. The 10 species of broad-footed marsupial mice live in rocky country with trees and undergrowth, especially near streams. They leave their shelters in rock crevices, hollow logs and caves at night, and are able to climb well.

A widespread species related to the broad-footed marsupial mice is Ingram's planigale, *Planigale ingrami*. It is the smallest of the living marsupials in Australia, being just under 4 inches (10 cm) long, of which nearly half is tail. The other planigales are also small, 2–4 inches (5–10 cm) long with 1¾–3½-inch (4.5–9-cm) tails. The flattened skulls of the planigales seem a clear adaptation to creeping through crevices in rock in the rocky and sandy areas they inhabit.

The mulgara, *Dasycercus cristicauda*, is one of the largest of the marsupial mice, being 5–8⅔ inches (12.5–22 cm) in head and body length with a 3–5-inch (7.5–12.5-cm) tail. It lives in deserts and has very short legs and a thick, well-furred tail. This contrasts with the jerboa marsupials, or kultarrs, which also live in the deserts of central Australia but have very long hind legs and a long, slender tail with a tuft of long hair at the tip. They spend the day in deep burrows or in hollow logs, coming out in the evening and remaining active until dawn.

Little predators

Marsupial mice are mainly predatory, probably eating any animal food small enough for them to tackle. The full list of their diet is not unlike that of shrews and they are known to eat insect

Marsupial mice are so named because some species have pouches in which they carry their young for the first 6 or 7 weeks of life. Pictured is Dasyuroides byrnei.

A fat-tailed dunnart eating a grasshopper. This species stores excess fat in its tail.

MARSUPIAL MICE

CLASS	**Mammalia**
ORDER	**Dasyrumorphia**
FAMILY	**Dasyuridae**
GENUS	**15, including broad-footed marsupial mice, *Antechinus*; and short-haired marsupial mice, *Murexia***
SPECIES	**55, including Macleay's marsupial mouse, *Antechinus stuartii*; Ingram's planigale, *Planigale ingrami*; and mulgara, *Dasycercus cristicauda***

ALTERNATIVE NAMES
Marsupial jerboa; tuan; planigale; dunnart; kultarr

WEIGHT
⅛–6 oz. (5–170 g); average: 3½ oz. (100 g)

LENGTH
Head and body: 2–8⅔ in. (5–22 cm); tail: 1¾–5 in. (4.5–12.5 cm)

DISTINCTIVE FEATURES
Resemble shrews; pointed muzzle; large eyes and ears; conspicuous vibrissae (whiskers); reddish brown to gray fur

DIET
Insects and their larvae, spiders, centipedes, small lizards, eggs and flower nectar

BREEDING
Breeding season: not known; number of young: up to 8; gestation period: 12–20 days (smaller species), 35–42 days (larger species)

LIFE SPAN
Up to 1½ years

HABITAT
Savanna, grassland, forest, scrub and bush

DISTRIBUTION
Australia, Tasmania and New Guinea

STATUS
Generally locally common; several species endangered or vulnerable

Marsupial mice

---

larvae and adult insects, including beetles, cockroaches and termites, as well as centipedes and spiders. They also take reptiles, for example small lizards, and eggs. The broad-footed marsupial mice also feed on flower nectar.

The large mulgara is highly valued because it preys upon the introduced house mouse, *Mus musculus*. One was seen to kill a house mouse with a single bite to the back of the neck. Then, starting at the tip of the snout, it methodically ate its way to the tip of the tail, turning the skin neatly inside-out as it went. Another species, which itself weighed only ¾ ounce (20 g), ate a whole ounce (28 g) of food in one night, comprising five large insect grubs and three small lizards, even the bones of the lizards being eaten.

Only some have pouches

Although we speak of these mammals as marsupial mice, not all have pouches. In some species the pouch consists of nothing more than a fold of skin surrounding the teats. In others it is a complete pouch, as in a kangaroo. In some marsupial mice, noticeably the jerboa marsupials, the pouch opens backward. Even when the pouch is incomplete, the babies remain attached to the teats for 6 or 7 weeks, the same length of time as they stay in the pouch in other species. Some of the marsupial mice may carry up to 8 young at a time. The young are independent after about 9–12 weeks.

Marsupial mice, even those that are widespread, seem unable to survive near human settlements where domestic cats are kept or have gone wild. Elsewhere they are preyed upon by snakes and owls and by their larger relatives, the marsupial cats, and the Tasmanian devil. Several species, including *Sminthopsis douglasi*, and one of the two species of brush-tailed marsupial mice, *Phascogale calwa*, are endangered. Others are threatened or vulnerable.

MARTEN

MARTENS ARE MEMBERS OF the same family of mammals as the badgers, otters, ferrets, weasels and ermines. They resemble large ferrets or weasels, with long, lithe bodies, short legs and pointed muzzles. There are several species in both the New World and the Old World.

The American marten, *Martes americana*, of North America, is about 20–23 inches (50–58 cm) long, with an 8-inch (20-cm) tail. It has light red to blackish fur and a buff- or orange-colored throat patch. It lives in forests, showing a preference for conifers, and ranges across most of the northern United States, Canada and Alaska. The fisher or pekan, *M. pennanti*, is larger than the American marten at 20–25½ inches (50–65 cm) long, and has a similar range. It is dark brown or black and lacks a throat patch. Six other species of fishers live in Europe and Asia.

The Eurasian pine marten, *M. martes*, is 18–23 inches (45–58 cm) long and weighs 1¾–4 pounds (0.8–1.8 kg); females are smaller than males. It has a bushy tail and a broad, triangular head. The coat is a rich brown, darker on the middle of the back and the legs. The underparts are paler brown, and on the throat there is a cream patch that may have an orange tinge in winter. It is the upper long guard hairs of the fur that give the coat color. Beneath these there is a short undercoat of reddish gray hairs tipped with yellow. The Eurasian pine marten still lives in the wilder parts of the British Isles, and extends across Europe and onto the plains of eastern Siberia. Farther east lives the closely related sable, *M. zibellina*. The beech or stone marten, *M. foina*, is found throughout continental Europe and from the Baltic Sea south to the Mediterranean and eastward to the Himalayas and Mongolia. The yellow-throated marten, *M. flavigula*, lives to the east of the beech marten.

Nimble tree climbers

Martens are excellent climbers, ascending and descending tree trunks headfirst and leaping through networks of thin branches with the ease of tree squirrels. If they fall, they land feet first, in the manner of a small cat.

In some places martens live on open rocky ground but even there they are only occasionally seen by humans. This is partly because they are

The Eurasian pine marten favors well-wooded areas. Marten dens are often found among pine trees, a fact that probably inspired their name.

Despite its name, the fisher does not catch live fish. Its diet comprises a range of animal food, including porcupines, rodents, birds, eggs and carrion.

CLASS	**Mammalia**
ORDER	**Carnivora**
FAMILY	**Mustelidae**
GENUS AND SPECIES	***Martes americana***

ALTERNATIVE NAMES
American pine marten; pine marten

WEIGHT
1¾–4 lb. (0.8–1.8 kg)

LENGTH
Head and body: 18–23 in. (50–58 cm); tail: 5¼–9¼ in. (13.5–24 cm)

DISTINCTIVE FEATURES
Long weasel-like body; catlike pointed ears; thick coat of variable color, from pale reddish to dark brown or blackish; cream or pale orange patch on throat and upper chest; long and bushy, foxlike tail

DIET
Mainly small mammals, including rabbits, jack rabbits, hares and mice; also birds, beetles and fruits

BREEDING
Age at first breeding: 1–2 years; breeding season: summer; number of young: 2 to 5; gestation period: 180–300 days including delayed implantation; breeding interval: about 1 year

LIFE SPAN
Up to 10 years

HABITAT
Forests, especially coniferous; also rocky grassland and steppe

DISTRIBUTION
Much of northern North America

STATUS
Population increasing; reintroduction programs in some areas

solitary and often nocturnal animals, but also because they are becoming increasingly rare in many areas.

Like all the members of their family, which includes the skunks, the American and Eurasian pine martens have special scent glands at the base of the tail. The secretions from these glands are used to mark their home range. Only skunks and the common polecat, *Mustela putorius*, use their secretions to deter enemies. The odor of the Eurasian pine marten is not objectionable to humans, and it was once called the sweet marten to distinguish it from the polecat.

Efficient predators

Eurasian pine martens chase squirrels through the trees. They also hunt rabbits, hares, rats, mice, voles, small birds and occasionally game birds, and will eat caterpillars, beetles and bees, as well as blackberries, bilberries and cherries. Pine martens eat slugs after rolling them under their paws to remove their slime, and dig up bees' nests in order to reach grubs and honey.

The American marten has a diet similar to that of the Eurasian pine marten, but the fisher hunts more frequently on the ground. The name of the latter is a misnomer as it probably feeds

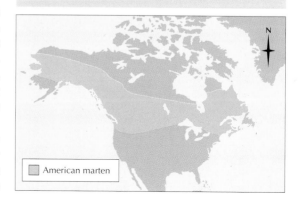

American marten

only on spent salmon or steals fish from traps and nets. The fisher catches beavers, American martens and smaller animals and kills porcupines by flipping them onto their backs. It has also been known to kill young deer that have foundered in deep snow.

Long gestation

Mating takes place in June or July but young Eurasian pine martens are usually not born until the following April as there is delayed implantation of the embryos. There are usually three to four young in each litter, although there may be any number from two to five. They may be born in a grass nest among rocks, in a hollow tree or in the old nest of a crow or squirrel. The young first come out of the nest when they are 2 months old and romp around nearby, producing a conspicuous playground of flattened grass and herbs. They are weaned at 6 or 7 weeks and leave their mother soon afterward. Female pine martens are mature when 1 year old, produce their first litter after 2 years and have been known to live for about 18 years in captivity.

The breeding habits of the American marten and fisher are similar, except that the American marten is weaned and becomes independent earlier. The female fisher leaves her litter in order to mate, then returns to them and rears them while the development of her new litter is delayed. Zoologists believe that other species of

martens behave in the same fashion, as the male does not stay with the female while she rears the young.

Trapped for the fur trade

In the Middle Ages martens were abundant in Europe, but they were subsequently trapped in such large numbers for their fur that they eventually became rare. They were also trapped in the cause of game preservation in Britain. The last marten to be killed near London was shot in Epping Forest in 1883. In Siberia the sable was all but exterminated for the sake of its rich fur. Just when the species seemed to be heading irreversibly for extinction, it was discovered that sable could be reared on ranches in the manner of mink. Since that time they have been bred not only for their fur but also for reintroduction into the places where they had been exterminated.

In North America a similar set of circumstances existed. Under the name of "American sable" or "marten," thousands of pelts were sold annually and marten populations were exterminated in many places. American martens do not breed readily in captivity and so repopulation proved to be problematic. Martens are very easy to catch, which is probably the primary reason for the serious reduction in their numbers. Limited protection on both sides of the Atlantic, however, has allowed the numbers of martens to recover in many places.

In common with other marten species, the American marten is a skillful climber. It is most often found in coniferous forests.

Index

Page numbers in *italics* refer to picture captions.
Index entries in **bold** refer to guidepost or biome and habitat articles.

Page numbers in *italics* refer to picture captions. Index entries in **bold** refer to guidepost or biome and habitat articles.

Page numbers in *italics* refer to picture captions. Index entries in **bold** refer to guidepost or biome and habitat articles.